Lecture Notes in Computer Science 12091

More information about this series at http://www.springer.com/series/7407

Bogumił Kamiński · Paweł Prałat ·
Przemysław Szufel (Eds.)

Algorithms and Models for the Web Graph

17th International Workshop, WAW 2020
Warsaw, Poland, September 21–22, 2020
Proceedings

 Springer

Editors
Bogumił Kamiński
SGH Warsaw School of Economics
Warsaw, Poland

Paweł Prałat
Ryerson University
Toronto, ON, Canada

Przemysław Szufel
SGH Warsaw School of Economics
Warsaw, Poland

ISSN 0302-9743   ISSN 1611-3349   (electronic)
Lecture Notes in Computer Science
ISBN 978-3-030-48477-4   ISBN 978-3-030-48478-1   (eBook)
https://doi.org/10.1007/978-3-030-48478-1

LNCS Sublibrary: SL1 – Theoretical Computer Science and General Issues

This Springer imprint is published by the registered company Springer Nature Switzerland AG
The registered company address is: Gewerbestrasse 11, 6330 Cham, Switzerland

Preface

The 17th Workshop on Algorithms and Models for the Web Graph (WAW 2020) was originally scheduled to take place at the SGH Warsaw School of Economics, Warsaw, Poland, during June 18–19, 2020. Unfortunately, because of the COVID-19 pandemic, the event was postponed and at the time of writing this text it is scheduled to take place during September 21–22, 2020.

This is an annual meeting, which is traditionally co-located with another, related, conference. WAW 2020 was planned to be co-located with the Workshop on Hypergraph Modelling. Co-location of the two workshops provides opportunities for researchers in two different but interrelated areas to interact and to exchange research ideas. We hope that WAW provides an effective venue for the dissemination of new results and for fostering research collaboration.

The World Wide Web has become part of our everyday life, and information retrieval and data mining on the Web are now of enormous practical interest. The algorithms supporting these activities combine the view of the Web as a text repository and as a graph, induced in various ways by links among pages, hosts, and users. The aim of the workshop was to further the understanding of graphs that arise from the Web and various user activities on the Web, and stimulate the development of high-performance algorithms and applications that exploit these graphs. The workshop gathered the researchers who are working on graph-theoretic and algorithmic aspects of related complex networks, including social networks, citation networks, biological networks, molecular networks, and other networks arising from the Internet.

This volume contains the papers accepted to WAW 2020. Each submission was carefully reviewed by the members of the Program Committee. Papers were submitted and reviewed using the EasyChair online system. The committee members decided to accept 12 papers.

April 2020

Bogumił Kamiński
Paweł Prałat
Przemysław Szufel

Organization

General Chairs

Andrei Z. Broder   Google, USA
Fan Chung Graham   UC San Diego, USA

Organizing Committee

Bogumił Kamiński   SGH Warsaw School of Economics, Poland
Paweł Prałat   Ryerson University, Canada
Przemysław Szufel   SGH Warsaw School of Economics, Poland

Sponsoring Institutions

The Polish National Agency for Academic Exchange (NAWA)
SGH Warsaw School of Economics
Ryerson University
Google
Moscow Institute of Physics and Technology
Internet Mathematics

Program Committee

Konstantin Avratchenkov   Inria, France
Mindaugas Bloznelis   Vilnius University, Lithuania
Paolo Boldi   University of Milan, Italy
Anthony Bonato   Ryerson University, Canada
Milan Bradonjic   Apprentice.io, USA
Fan Chung Graham   UC San Diego, USA
Collin Cooper   King's College London, UK
Andrzej Dudek   Western Michigan University, USA
Alan Frieze   Carnegie Mellon University, USA
Jeannette Janssen   Dalhousie University, Canada
Cliff Joslyn   Pacific Northwest National Laboratory, USA
Bogumil Kaminski   SGH Warsaw School of Economics, Poland
Ravi Kumar   Google, USA
Marc Lelarge   Inria, France
Stefano Leonardi   Sapienza University of Rome, Italy
Lasse Leskela   Aalto University, Finland
Nelly Litvak   University of Twente, The Netherlands
Michael Mahoney   UC Berkeley, USA

Contents

Hypergraph Analytics of Domain Name System Relationships . . . .　1
　Cliff A. Joslyn, Sinan Aksoy, Dustin Arendt, Jesun Firoz, Louis Jenkins,
　Brenda Praggastis, Emilie Purvine, and Marcin Zalewski

Global Graph Curvature. . . . .　16
　Liudmila Prokhorenkova, Egor Samosvat, and Pim van der Hoorn

Information Diffusion in Complex Networks: A Model Based
on Hypergraphs and Its Analysis. . . . .　36
　Alessia Antelmi, Gennaro Cordasco, Carmine Spagnuolo,
　and Przemysław Szufel

A Scalable Unsupervised Framework for Comparing Graph Embeddings . . . .　52
　Bogumił Kamiński, Paweł Prałat, and François Théberge

Assortativity and Bidegree Distributions on Bernoulli Random
Graph Superpositions. . . . .　68
　Mindaugas Bloznelis, Joona Karjalainen, and Lasse Leskelä

Clustering Coefficient of a Preferred Attachment Affiliation Network . . . . . .　82
　Daumilas Ardickas and Mindaugas Bloznelis

Transience Versus Recurrence for Scale-Free Spatial Networks. . . . .　96
　Peter Gracar, Markus Heydenreich, Christian Mönch,
　and Peter Mörters

The Iterated Local Directed Transitivity Model for Social Networks . . . .　111
　Anthony Bonato, Daniel W. Cranston, Melissa A. Huggan,
　Trent Marbach, and Raja Mutharasan

A Note on the Conductance of the Binomial Random Intersection Graph. . . .　124
　Katarzyna Rybarczyk, Mindaugas Bloznelis, and Jerzy Jaworski

Iterated Global Models for Complex Networks . . . .　135
　Anthony Bonato and Erin Meger

A Robust Method for Statistical Testing of Empirical
Power-Law Distributions . . . .　145
　Davide Garbarino, Veronica Tozzo, Andrea Vian, and Annalisa Barla

Community Structures in Information Networks for a Discrete
Agent Population 158
   Peter Marbach

Author Index 173

Hypergraph Analytics of Domain Name System Relationships

Cliff A. Joslyn[1][(✉)], Sinan Aksoy[2], Dustin Arendt[2], Jesun Firoz[1],
Louis Jenkins[3], Brenda Praggastis[1], Emilie Purvine[1], and Marcin Zalewski[4]

[1] Pacific Northwest National Laboratory, Seattle, WA, USA
cliff.joslyn@pnnl.gov
[2] Pacific Northwest National Laboratory, Richland, WA, USA
[3] University of Rochester, Rochester, NY, USA
[4] NVIDIA, Santa Clara, CA, USA

Abstract. We report on the use of novel mathematical methods in hypergraph analytics over a large quantity of DNS data. Hypergraphs generalize graphs, as used in network science, to better model complex multiway relations in cyber data. Specifically, casting DNS data from Georgia Tech's ActiveDNS repository as hypergraphs allows us to fully represent the interactions between *collections* of domains and IP addresses. To facilitate large-scale analytics, we fielded an analytical pipeline of two capabilities: HyperNetX (HNX) is a Python package for the exploration and visualization of hypergraphs; while on the backend, the Chapel HyperGraph Library (CHGL) is a library for high performance hypergraph analytics written in the exascale programming language Chapel. CHGL was used to process gigascale DNS data, performing compute-intensive calculations for data reduction and segmentation. Identified portions are then sent to HNX for both exploratory analysis and knowledge discovery targeting known tactics, techniques, and procedures.

Keywords: Hypergraphs · DNS · High performance computing · Chapel

1 Introduction

Many problems in data analytics involve rich interactions amongst multiple entities, for which graph representations are commonly used. High order (high dimensional) interactions abound in cyber and social networks, and can only be represented in graphs as highly inefficiently coded, "reified" labeled subgraphs. Lacking multi-dimensional relations, it is hard to address questions of "community interaction" in graphs: how is a collection of entities A connected to another collection B through chains of other communities?; where does a particular community stand in relation to other communities in its neighborhood?

© Springer Nature Switzerland AG 2020
B. Kamiński et al. (Eds.): WAW 2020, LNCS 12091, pp. 1–15, 2020.
https://doi.org/10.1007/978-3-030-48478-1_1

Hypergraphs [4] are generalizations of graphs which allow edges to connect any number of vertices. Hypergraph methods are well known in discrete mathematics, and are closely related to important objects in data science such as bipartite graphs, set systems, partial orders, finite topologies, and especially graphs proper, which they directly generalize (every graph is a 2-uniform hypergraph). In HPC, hypergraph-partitioning methods help enable parallel matrix computations [8], and have applications in VLSI [13]. In the network science literature, researchers have devised several path and motif-based hypergraph data analytics (albeit fewer than their graph counterparts), such as in clustering coefficients [15] and centrality metrics [9].

Complex data commonly analyzed using network science methods, and especially including cyber data, often contain multi-way interactions. But while they thus present naturally as hypergraphs, still hypergraph treatments are very unusual compared to graph representations of the same data. This is due at least to the greater mathematical, conceptual, and computational complexity of hypergraph methods, since as data objects, hypergraphs scale as $O(2^n)$ in the number of vertices n, as opposed to $O(n^2)$ for graphs. In the face of this, complex data are typically collapsed or are simplified to graphs to ease analysis.

We are accepting the challenge of the complexity of hypergraphs in order to gain the formal clarity and support for analysis of complex data they provide. A substantial high-performance computing (HPC) component is thus necessary, despite hypergraph analytics not receiving much attention in the software engineering community at large, and the HPC community in particular. We thus pursue a two-fold approach to developing our methods:

1. The **C**hapel **H**ypergraph **L**ibrary (CHGL, https://github.com/pnnl/chgl) [12] is a library for hypergraph computation in the emerging Chapel programming language [6,7], for HPC hypergraph processing, large scale analysis, and data segmentation.
2. We explore single hypergraphs or collections of hypergraphs using **HyperNetX** (HNX, https://github.com/pnnl/HyperNetX), a Python library for exploratory data analytics and visualization of hypergraphs.

In our work, CHGL and HNX are two stages of an analytical pipeline: CHGL provides a highly abstract interface for implementation of HPC hypergraph algorithms over large data, identifying segments and subsets which can then be passed to HNX for more detailed analysis.

In this paper we first introduce the foundations of hypergraph mathematics and hypernetwork science in the context of our CHGL and HNX capabilities. We then describe the DNS data set, selections of the ActiveDNS data sets from the Georgia Institute of Technology [1]. We then describe CHGL, before going on to describe the results of our demonstration analyses. These include both basic global statistics like degree and edge size distributions, as well as exploratory discovery of small components involving motif mining and computation of simple hypergraph metrics to discover outliers.

2   Hypergraph Analytics

An undirected **hypergraph** is a pair $\mathcal{H} = \langle V, \mathcal{E}\rangle$ with V a finite, non-empty set of **vertices**, and $\mathcal{E}$ a non-empty multiset of **hyperedges** $e \in \mathcal{E}$ (or just "edges" when clear), where $\forall e \in \mathcal{E}, e \subseteq V$. Hypergraphs can be represented in many forms, two of which are shown in Fig. 1 for a small example $\mathcal{H}$ with $V = \{1, 2, \ldots, 9\}$, representing $|V| = 9$ IP addresses.[1] On the left is an Euler diagram showing each of eight hyperedges $A, B, \ldots, H$, representing domains, as a "lasso" around its vertices. On the right is a $V \times \mathcal{E}$ incidence matrix I, where a non-null $\langle v, e\rangle \in I$ cell indicates that $v \in e$ for some $v \in V, e \in \mathcal{E}$.

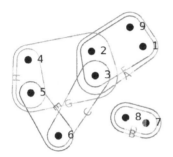

	A	B	C	D	E	F	G	H
1	X					X		
2	X					X	X	
3	X		X			X	X	
4							X	X
5					X		X	X
6			X		X			
7		X		X				
8		X		X				
9	X					X		

Fig. 1. (Left) An Euler diagram of an example hypergraph $\mathcal{H}$. (Right) Its incidence matrix I.

We call each hyperedge $e \in \mathcal{E}$ an s-edge where $s = |e|$. Thus all graphs are hypergraphs, in that all graph edges are 2-edges, for example $H = \{4, 5\}$, saying that the domain H has two IPs 4 and 5. But $F = \{1, 2, 3, 9\}$ is a 4-edge, with domain F having those four IPs. Where each column of the incidence matrix of a graph has exactly two cells, those of hypergraphs are unrestricted.

Our research group is pursuing hypergraph analytics as an analog to graph analytics [14]. While our development is consistent with others in the literature [9,16], our notation and concepts are somewhat distinct. For a more comprehensive development see [2].

We say that two edges $e, f \in \mathcal{E}$ are s-**adjacent** if $|e \cap f| \geq s$ for $s \geq 1$. An s-**star** is a set of edges $\mathcal{F} \subseteq \mathcal{E}$ sharing exactly a common intersection $f \subseteq V$, with $|f| \geq s$, so that $\forall e_i, e_j \in \mathcal{F}$ we have $e_i \cap e_j = f$. An s-**path** is a sequence of edges $\langle e_0, e_1, \ldots, e_n\rangle$ such that each e_{i-1}, e_i are s-adjacent for $1 \leq i \leq n$; and an s-**component** is a maximal collection of edges any pair of which is connected by an s-path. The s-**diameter** of an s-component is the length of its longest shortest s-path. Comparing again to graphs, graph paths are all 1-paths, and graph

---

[1] $\mathcal{H}$ can also be represented as a bipartite graph on the disjoint union $V \sqcup \mathcal{E}$, with each component a distinct part.

components all 1-components. Our example has two 1-components (shown obviously), but also four 2-components (listed edge-wise) $\{A, F, G, H\}, \{B, D\}, \{C\}$ and $\{E\}$. Its 3- and 4-components are each single edges of size larger than 3 or 4 (respectively), and it has no 5 or higher components.

Given a hypergraph $\mathcal{H}$, it is possible to construct smaller representations which capture important properties:

- Note that in our example, the edges $A = F$ and $B = D$, and the vertices $1 = 9$ and $7 = 8$, are equivalent, represented as duplicate columns and rows in I respectively. **Collapsing** is the process of combining these and replacing them with a representative, while also possibly maintaining a multiplicity count to be used for a weighting. The edges $\mathcal{E}$ are hereby transformed from a multiset to a set.
- Additionally, note that after collapsing, the smaller 1-component becomes an **isolated singleton**, effectively a collection of non-interacting vertices, or a diagonal block in I. These are especially common in DNS data. Pre-collapse, an isolated singleton would indicate the normal, *uninteresting* behavior in DNS where a single IP is associated with a single domain, and *vice versa*. But post-collapse, they indicate a collection of IPs and domains which are universally associated only with themselves, effectively forming a set of domain and IP aliases. In this work, these are counted and pruned, but in the future they could be attended to with respect to their multiplicities.
- Finally, note that $H \subset G$ is an **included** edge. Non-included edges are called **toplexes**, and not only is the collection of toplexes much smaller than $\mathcal{H}$, but it is sufficient to derive some hypergraph information, for example s-components.

Table 1 shows some important statistics for our example, first for the initial hypergraph, then after collapsing, and finally after removing isolated singletons from the collapsed hypergraph. For hypergraph data, a vastly high or low aspect ratio can indicate difficulty in analysis. Note that as reductions commence, the number of vertices, edges, and cells reduces, while density increases. Finally, Fig. 2 shows the distribution of node degree (# edges per node) and edge size.

Table 1. Basic hypergraph statistics for our example.

	Initial	Collapsed	Non-singleton components
$|V|$	9	7	6
$|E|$	8	6	5
Aspect ratio	1.125	1.167	1.200
# Cells	23	14	13
Density	0.319	0.333	0.433

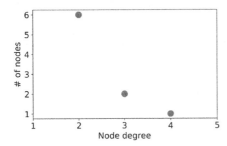 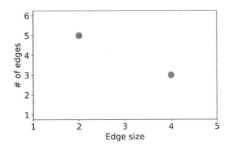

Fig. 2. (Left) Distribution of node degree (# edges per node) in our example. (Right) Distribution of edge size s.

In our pipeline the segmentation steps of collapsing, removing isolated singletons, and computing s-components are all performed using CHGL, as are node degree, edge size, and s-component size distributions. Subsequent exploration of the structures found within the components themselves, e.g., identification of stars and computation of diameters, are done via HNX. HNX builds on the popular library NetworkX [10], which offers metrics and algorithms for the analysis of graph data. Euler diagram visualizations that appear in this paper are provided directly by the HNX package.

3   Hypergraph Representations of DNS Data

The Domain Name System (DNS) provides a decentralized service to translate from the domain names that humans keep track of (e.g., www.google.com) to IP addresses that computers require to communicate. Perhaps somewhat counter-intuitively, DNS data present naturally as a hypergraph, in being a many-many relationship between domains and IPs. While typically this relationship is one-to-one, with each domain uniquely identifying a single IP address and *vice versa*, there are a number of circumstances which can violate this:

- Some domains have aliases so that multiple domains (e.g., misspellings) resolve to the same IP address.
- There are large hosting services where one IP serves multiple websites.
- Some domains are used so frequently that they must be duplicated across hosts and therefore map to multiple IPs.
- IP addresses are randomly reassigned within some small IP block so the same domain may map to multiple IP addresses when queried over time.

In order to explore large volumes of DNS mappings we used ActiveDNS (ADNS), a data set maintained by the Astrolavos Lab at Georgia Institute of Technology (https://activednsproject.org). ADNS submits daily DNS lookups for popular zones (e.g., .com, .net, .org) and lists of domain names. The data is stored in Avro format (https://avro.apache.org) which provides structured records for each DNS lookup in a compressed binary file. Each record contains information including: query date, lookup input (often a domain name), data returned by a DNS server (often a list of IP addresses), and IP addresses of the DNS servers that answered the query. DNS records are also typed according to different properties (recorded as the `qtype` field in ADNS) such as the format of the data and to indicate its intended use. This initial analysis accepted any reasonable pairing of domain name IP address and did not restrict to any particular `qtypes`. Future work will restrict to `qtype = 1`, which map hostnames to an IPv4 address of the host.

Our group acquired data from the time period April 24–May 29, 2018, and in this paper we focus on the single day of April 26, 2018. This day consists of 1,200 Avro files with each file containing on average 900 K records. There was some data cleaning necessary to remove records with empty lookup input or empty returned data. Additionally we removed any records in which the lookup input was an IP address or the returned data was a domain name. After cleaning, each file was reduced to approximately 180 K records.

We structured these DNS data as a hypergraph on a vertex set V of IPs and edge set $\mathcal{E}$ of domains. Thus our hypergraphs $\mathcal{H}$ coded each domain as a collection of its IPs. We show results of our anlaysis below in Sect. 5, including global statistics and the results of targeted exploration.

4 Chapel Hypergraph Library (CHGL)

The Chapel HyperGraph Library (CHGL) [12] is a prototype exascale library written in Chapel [6, 7] that brings generation, representation, and computation of hypergraphs to the world of high performance computing (HPC). Thanks to Chapel, CHGL provides scalability in both shared memory and distributed memory contexts.

In most cases, data underlying a hypergraph is more complex than CHGL's internal representation of vertices and hyperedges as consecutive integers. In such situations, a hash table that maps user-defined generic properties to the consecutive identifiers of vertices and hyperedges is used for translation. The properties are embedded in the internal representation of the hyperedges and vertices, allowing $O(1)$ bidirectional lookup as well as locality when iterating over the graph, shared-memory and distributed alike.

CHGL performs *segmentation*, or reduction, of the data in multiple highly-parallel phases. Segmentation reduces both the size of the graph to one that HNX can process in a reasonable amount of time and the computational workload on CHGL when computing metrics. Proper care is taken to ensure that references to the collapsed hyperedges and vertices are taken forward to the hyperedge or vertex that they collapsed into, and that all references to removed hyperedges and vertices are removed. This is performed in linear time and applies to both the graph and property map.

To prune away redundant entities, which is generally useful for computation, hyperedges and nodes are placed into equivalence classes through the process of collapsing described in Sect. 2. All but one arbitrarily chosen representative is removed from the graph. Determining the equivalence class of a vertex or hyperedge can be done by using a set or hash table, and can be performed in $O(|V|)$ or $O(|V|\log|V|)$ time, depending on the data structure used. In practice, the time complexity is often linear or quasilinear, but in the worst-case scenario when the hypergraph is fully connected, the time complexity is $O(|V|^2)$ or $O(|V|^2\log|V|)$.

Isolated singletons, as described in Sect. 2, tend to be uninteresting. After collapsing, these are pruned away in a straightforward manner.

We implemented computation of s-components using a parallel search method, where we iterate over edges in parallel, and every edge begins an independent search. The s-neighbors of an edge are marked with the component number originating from the initial edge. The component number is taken from a global atomic counter at the beginning of every parallel search. 1-Components are implemented by simply traversing the edges by following included vertices (edge to vertex to neighbor edge), but 2-components and higher require an implementation with set intersections to check the cardinalities of adjacencies. This implementation is well suited for a large number of small components because most components end up being searched by a single task. The best case scenario complexity of the parallel search algorithm is linear, and the worst is quadratic if the maximum number of component collisions occur. The average complexity in our case is close to linear since the DNS data has a large number of small components, and most components are handled by a single task.

Obtaining the vertex degree and edge cardinality distributions is simple and intuitive in CHGL, thanks to Chapel's high-level abstractions. This particular operation is short enough that it can be presented in full in Fig. 3. We compute these both pre- and post-collapsing.

```
1   // Find largest degree of all vertices in the hypergraph
2   var N = max reduce [v in graph.getVertices()] graph.degree(v);
3   var degreeDist : [1..N] int;
4   forall v in graph.getVertices() with (+ reduce degreeDist) {
5     degreeDist[graph.degree(v)] += 1;
6   }
```

Fig. 3. Obtaining the vertex degree distribution in CHGL.

Finally, the s-component size distributions were computed, recording the number of nodes and edges in each s-component and how many s-components have each size. This allows us to understand how nodes and edges are distributed, e.g., is there one giant component and a few small components or are component sizes more uniformly distributed.

5   Computational Results

We ran experiments on one of the compute nodes of an Infiniband cluster, each equipped with a 20-core Intel Xeon processor and 132 GB memory. All cores were involved in the experiments. CHGL v0.1.3 was compiled against Chapel pre-release version 1.18.0 with `--fast` flag to enable all compiler optimizations.

Execution times of the stages of the CHGL DNS processing pipeline are shown on the left side of Fig. 4. s-component computation dominates the execution time for 128 or more files. The s-components are reused when computing the s-component size distributions, leading to them taking significantly less time. Collapsing duplicates and removing isolated components scale linearly, as is expected for their time complexity. The hypergraph is constructed in about the same amount of time it takes to collapse it, showing that processing DNS data is mostly compute-bound.

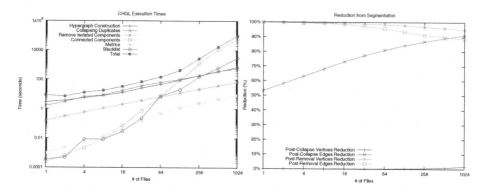

Fig. 4. (Left) Execution times (log-log scale). (Right) Effectiveness of reduction from segmentation.

The purpose of segmentation is to reduce the size of the graph while also maintaining the data that is of interest. The right side of Fig. 4 shows the compression as a result of performing segmentation. Collapsing of duplicate edges results in the most compression, reducing the graph from 55% at one file to over 90% at 1024 files, which can be expected to improve further when more data is processed. Removing isolated components results in less compression as data size increases, likely due to the premature marking of components as isolated prior to having all of the data. Perhaps with larger amounts of data, there will be a convergence to a stable number of isolated components in the entirety of the DNS network. Note that there are very few duplicate IP addresses on smaller samples, but that may change as more data is processed; nonetheless, collapsing duplicate vertices may be unnecessary and can possibly save some time.

Above we reported on scaling of loading and compute time using CHGL on varying numbers of ActiveDNS files, from 1 to 1,024. Here we report on analysis of the hypergraph built from one full day, April 26, 2018, comprising 1,200 files. See Table 2 for basic count statistics.

The node degree and edge size distributions are shown in Figs. 6a and 6c. Except for the small increase around $x = 10^2$ the node degree distribution looks

Table 2. Basic hypergraph statistics for ActiveDNS data for April 26, 2018.

	Initial	Collapsed	Non-singleton components
$|V|$	10.6M	10.3M	557K
$|E|$	131.2M	11.0M	1.2M
Aspect ratio	0.081	0.941	0.460
# cells	157.4M	25.7M	15.9M
Density	1.14 E-7	2.26 E-7	2.35 E-5

like a power law or heavy tailed distribution typical in real-world graphs [3]. The degree distribution has a general decreasing tendency from $x = 1$ to $x = 70$, it increases by roughly 1,000 through $x = 80$, and then returns to the downward trend. We do not know why this occurs, but it is possible that it could be an artifact of DNS server configuration practices. Edge size distribution also seems to be heavy-tailed although somewhat more noisy for low edge sizes than the degree distribution.

See the second column in Table 2 for the simple count statistics of the collapsed hypergraph. Notice that collapsing resulted in a much more square incidence matrix since only 2% of nodes were collapsed while 92% of edges were collapsed. The number of cells in the collapsed hypergraph incidence matrix is now reduced to 16% of the full hypergraph.

The distributions of node and edge duplicate counts are shown in Fig. 5. Notice that the distribution of duplicate edge counts has a similar shape as the node degree distribution of the original hypergraph with a slight increase around $x = 10^2$. After seeing this it is possible that the nodes which had degree around 70–80, where this increase occurs, were actually in many duplicate edges which are now collapsed. The node degree distribution for the collapsed hypergraph found in Fig. 6b further supports this hypothesis since the increase around $x = 10^2$ in the node degree distribution is absent.

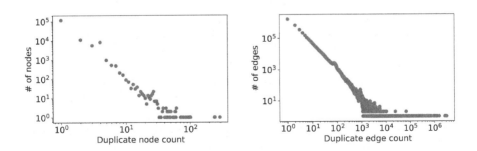

Fig. 5. Distribution of duplicate node counts (top) and edge counts (bottom).

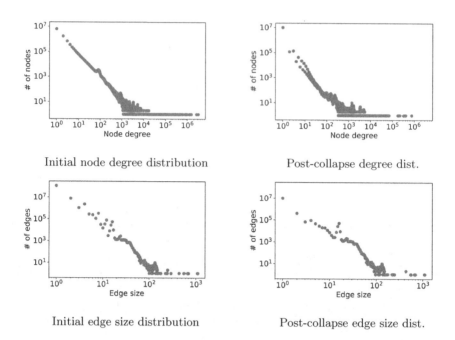

Initial node degree distribution   Post-collapse degree dist.

Initial edge size distribution   Post-collapse edge size dist.

Fig. 6. Node degree and edge size distributions, on a log-log scale, for April 26, 2018 DNS hypergraph. The x and y axes are the same across both node plots and across both edge plots to illustrate the changes through the collapsing procedure.

The edge size distribution post collapse is shown in Fig. 6d. This distribution is very similar to that of the original hypergraph, although it appears less noisy up through approximately $x = 20$. This is not surprising since there were not many duplicate nodes removed, so edges that remained likely stayed close to their original size.

After collapsing duplicate nodes and edges we removed all 9,784,763 isolated singleton edges, or 89% of all remaining edges. The only differences between the collapsed hypergraph and the hypergraph after removal of isolated singleton components is the number of degree 1 nodes and the number of size 1 edges. Therefore, we omit the final node degree and edge size distributions since they are identical to the post-collapse distributions except for the points at $x = 1$.

Comparing the pre-collapse (left), post-collapse (right), and post-removal distributions (not pictured) in Fig. 6, we observe that hypergraph collapsing and removal significantly alters the shape of degree and edge size distributions. In addition to the qualitative differences apparent from the plots, these differences can also be quantified using the Kolmogorov-Smirnov (KS) distance metric, a normalized statistic between 0 and 1 in which larger values indicate greater degree distribution dissimilarity. In the case of the degree distributions (top row), KS distance suggests the pre-collapsing hypergraph differs significantly from the post-collapse and post-removal degree distributions, with KS values of 0.36 and

0.34, respectively. In the case of the edge-size distributions (bottom row), the most pronounced difference is between the pre-collapsing and post-removal edge size distribution, with a KS value of 0.60. Here, the large KS distance reflects the dramatic changes at the head of the distribution, where the number of 1-edges decreases from 118 million to 369 thousand.

6 Analytical Results

The next step toward finding interesting subgraphs within the single day of ActiveDNS data was to compute and explore s-components. CHGL computed s-components of the hypergraph post-collapse and post-removal of isolated singletons for $s = 1, 2, 3$. Before exploring these components themselves we report the distribution of component sizes (both node and edge counts) which are found in Fig. 7. As s increases the shapes of these distributions do not change much but they do tend to skew more toward smaller components and the distribution flattens slightly. This is required since every s-component is contained within some s'-component for $s' < s$: as s increases components can only decrease in size. These distributions also show that while there are some very large s-components the majority are very small. Additionally, we see that the notion of a "giant component" is much more prevalent in the set of 1-components than for $s = 2$ or 3. Indeed, as s increases the largest component breaks up and the jump between the largest component and second largest becomes smaller.

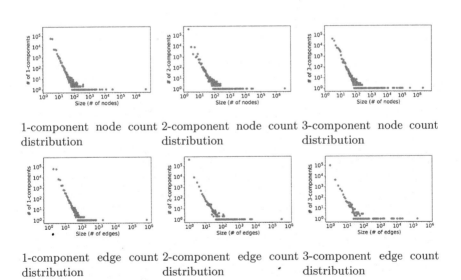

1-component node count distribution

2-component node count distribution

3-component node count distribution

1-component edge count distribution

2-component edge count distribution

3-component edge count distribution

Fig. 7. Node and edge count distributions, on a log-log scale, for s-components within simplified April 26, 2018 DNS hypergraph. The x and y axes are the same across all three node count plots and across all three edge count plots to illustrate the changes as s increases.

Once the hypergraphs were segmented into *s*-components by CHGL we proceeded to do exploratory analysis using HNX. In particular, we looked for:

– Occurrences of 1-stars within the 1-components, and
– *s*-components with maximum *s*-diameter for $s = 2, 3$.

Recall that a 1-star is a small hypergraph in which all edges pairwise intersect in one node, and that one node is the same across all pairwise intersections. The simplest 1-star has all edges of size 2, see Fig. 9a for an example of this case. In our DNS use case a star is a collection of domains which all share exactly one IP address but each also have their own separate IP address(es). These are consistent with the behavior of content delivery networks (CDN), geographically distributed networks of servers with the goal of quickly and reliably serving up content to a variety of users, which could explain the existence of stars with a diverse set of IP addresses since a consideration for IP assignment is geographic location. Star motifs are also consistent with DNS sinkholes and domain hosting services.

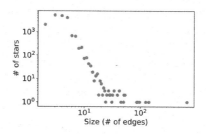

Fig. 8. Distribution of star sizes (# of edges).

We searched the 1-components for 1-stars and looked for size outliers. The distribution of number of edges per star is shown in Fig. 8. We can see that there is one notable outlier, a star with 701 edges and 642 toplexes. The domain names within this star appear to be mostly randomly generated within the .com and .net zones (e.g., `twlwta.com`, `comgslklpa.net`) and the common IP address within all domains is 17.17.17.17. A WHOIS search finds that this IP address is within the network range of Apple, Inc. The other 642 IPs present in this star come from 640 distinct of/16 ranges. This is consistent with "DNS sinkhole" behavior where traffic to a variety of (potentially malicious) domains is redirected to a benign location [5]. And later (i.e., not on April 26) DNS searches for a sample of domains within this star have a Start of Authority (SOA) record with "sinkhole root@sinkhole" as the name and contact for the server.

Unlike this largest star which had IP addresses in many different ranges, smaller stars such as the one shown in Fig. 9a tend to have all IPs and domains within the same, or a relatively small set of, ranges and organizations. In this small example WHOIS lookups indicate that the central IP address is

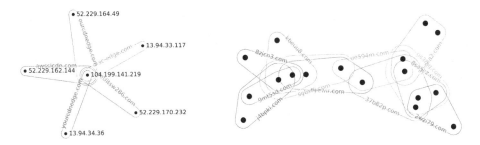

Fig. 9. (a) A small star seen in the ActiveDNS data. (b) The 2-component with largest 2-diameter.

from Google Cloud whereas the leaves are from Microsoft Corporation. All five domains are registered through the hosting site GoDaddy.com.

To discover interesting 2-components (resp. 3-components) we calculated 2-diameters (resp. 3-diameters) of each of the components and look more closely at those with maximal diameter. In the case of the 2-components the maximum 2-diameter is 6 and there is only one 2-component with that 2-diameter, shown in Fig. 9b. The IP addresses in this component all belong to the IP range 103.86.122.0/24 and the domains are registered to GMO INTERNET, INC according to WHOIS records. Moreover, current DNS queries for most of these domains at a later date resolve to IPs in the range 103.86.123.0/24 and have a time to live of only 120 s. This pattern of quickly changing of IP address is consistent with the *fast flux* DNS technique which can be used by botnets to hide malicious content delivery sites and make networks of malware more difficult to discover [11].

The large diameter 3-components tell different stories. The maximum 3-diameter is 3 and there are four 3-components with this 3-diameter. One has only one toplex with six sub-edges. Two others are fairly simple and, like the large 2-diameter 2-component, are somewhat chain-like tracing out a long path. The fourth is quite large with 70 nodes, 189 edges, and all IPs belonging to an IP range from Amazon Technologies Inc.

7   Conclusions and Future Work

While our research group has been developing hypergraph methods and mathematics over a moderate period, this paper reflects the first application of CHGL to cyber data.

The current approach is limited in a number of ways. First, ActiveDNS records data from DNS lookups on a daily basis (or perhaps multiple times per day), but it does not do continual monitoring. This discrete sampling may mean that the pipeline misses patterns that would normally be seen in a more continuous approach. Additionally, the current analysis is for a single day, and extending to multiple days in the current architecture will exacerbate issues with memory bounds. This might be mitigated using a theory of dynamic hypergraphs (much like that of dynamic graphs) to understand the time-evolution of DNS or similar data.

Additionally, certain DNS relationships are ignored, such as recursive DNS records where one domain resolves not to an IP address but to another domain name. This would require more complicated mathematics than just hypergraphs, likely cell complexes or partial orders, which we have started to consider in our research but not yet in our analysis. We also ignore other pieces of metadata like the authority IP addresses (those servers which answered the DNS request).

Additional future work includes:

- We are extending our prior theoretical work [2,14] to a full consideration of the mathematical foundations of hypergraphs for data science, including spectral approaches and consideration of multiplicity weightings.
- A range of hypernetwork methods generalizing network science centrality, connectivity, clustering coefficients, etc. are available [2].
- Also central to our approach is the consideration of hypergraphs as multidimensional objects, and thus inherently available for topological applications, including homology measurement for identification of loops and potential gaps in the underlying data.
- CHGL is also under active development to include topology, homology measures, a proper graph library, and a distributed data model.
- Finally, application and data analysis continues, including DNS, additional cyber data beyond DNS, and additional application domains including computational biology and social hypernetworks.

Acknowledgements. This work was partially funded by a US Department of Energy Computational Science Graduate Fellowship (grant DE-SC0020347).

This work was also partially funded under the High Performance Data Analytics (HPDA) program at the Department of Energy's Pacific Northwest National Laboratory. Pacific Northwest National Laboratory is operated by Battelle Memorial Institute under Contract DE-ACO6-76RL01830.

Special thanks to William Nickless for helpful conversations surrounding the DNS analysis and interpretation.

References

1. Active DNS project. https://activednsproject.org/. Accessed 26 Nov 2019
2. Aksoy, S.G., Joslyn, C., Marrero, C.O., Praggastis, B., Purvine, E.: Hypernetwork science via high-order hypergraph walks. arXiv preprint arXiv:1906.11295 (2019, Submitted)
3. Barabási, A.-L., Bonabeau, E.: Scale-free networks. Sci. Am. **288**(5), 60–69 (2003)
4. Berge, C., Minieka, E.: Graphs and Hypergraphs. North-Holland, Amsterdam (1973)
5. Guy Bruneau. DNS Sinkhole. https://www.sans.org/reading-room/whitepapers/dns/dns-sinkhole-33523

6. Chamberlain, B.L., Callahan, D., Zima, H.P.: Parallel programmability and the chapel language. Int. J. High Perform. Comput. Appl. **21**(3), 291–312 (2007)

7. Chamberlain, B.L., et al.: Chapel comes of age: Making scalable programming productive. Cray Users Group (2018)

8. Devine, K.D., Boman, E.G., Heaphy, R.T., Bisseling, R.H., Catalyurek, U.V.: Parallel hypergraph partitioning for scientific computing. In: Proceedings 20th IEEE International Parallel & Distributed Processing Symposium. IEEE (2006)

9. Estrada, E., Rodríguez-Velázquez, J.A.: Subgraph centrality and clustering in complex hyper-networks. Phys. A **364**, 581–594 (2006)

10. Hagberg, A.A., Schult, D.A., Swart, P.J.: Exploring network structure, dynamics, and function using networkx. In: Varoquaux, G., Vaught, T., Millman, J. (eds.) Proceedings of the 7th Python in Science Conference, Pasadena, CA USA, pp. 11–15 (2008)

11. Riden, J.: How fast-flux service networks work. http://www.honeynet.org/node/ 132. Accessed 26 Nov 2018

12. Jenkins, L.P., et al.: Chapel hypergraph library (CHGL). In: 2018 IEEE High Performance Extreme Computing Conference (HPEC 2018) (2018)

13. Karypis, G., Kumar, V.: Multilevel k-way hypergraph partitioning. VLSI Des. **11**(3), 285–300 (2000)

14. Purvine, E., Aksoy, S., Joslyn, C., Nowak, K., Praggastis, B., Robinson, M.: A topological approach to representational data models. In: Yamamoto, S., Mori, H. (eds.) HIMI 2018. LNCS, vol. 10904, pp. 90–109. Springer, Cham (2018). https:// doi.org/10.1007/978-3-319-92043-6_8

15. Robins, G., Alexander, M.: Small worlds among interlocking directors: network structure and distance in bipartite graphs. Comput. Math. Organ. Theory **10**(1), 69–94 (2004)

16. Wang, J., Lee, T.T.: Paths and cycles of hypergraphs. Sci. China, Ser. A Math. **42**(1), 1–12 (1999)

Global Graph Curvature

Liudmila Prokhorenkova[1,2(✉)], Egor Samosvat[1], and Pim van der Hoorn[3]

[1] Yandex, Moscow, Russia
ostroumova-la@yandex.ru
[2] Moscow Institute of Physics and Technology, Moscow, Russia
[3] Eindhoven University of Technology, Eindhoven, The Netherlands

Abstract. Recently, non-Euclidean spaces became popular for embedding structured data. However, determining suitable geometry and, in particular, curvature for a given dataset is still an open problem. In this paper, we define a notion of *global graph curvature*, specifically catered to the problem of embedding graphs. We theoretically analyze this value and show that the optimal curvature essentially depends on the dimensionality of the embedding space and loss function one aims to minimize via embedding. We also review existing notions of local curvature (e.g., Ollivier-Ricci curvature) and conduct a theoretical analysis of their properties. In particular, we demonstrate that the global curvature differs significantly from the aggregations of local ones. Thus, the proposed measure is non-trivial and it requires new empirical estimators as well as separate theoretical analysis.

Keywords: Graph embedding · Curvature · Non-Euclidean spaces

1 Introduction

Representation learning is an important tool for learning from structured data such as graphs or texts [4,10,16]. State-of-the-art algorithms typically use Euclidean space for embedding. Recently, however, it was found that hyperbolic spaces demonstrate superior performance for various tasks [13,17], while in some cases spherical spaces can be useful [9]. A key characteristic classifying the above-mentioned spaces is curvature, which is negative for hyperbolic spaces, zero for Euclidean, and positive for spherical spaces. These findings, therefore, show that certain graphs are better represented in spaces with non-zero curvature. While some methods simply fix the curvature (e.g., -1 for hyperbolic space) and then find the optimal embedding of the graph in the corresponding space [13], others try to learn the right curvature and embedding simultaneously [5].

In this paper, we consider the problem of determining a graph curvature suitable for embedding. We first introduce a concept of *global* graph curvature, which depends on both the dimension and loss function used for the embedding.

Supported by the Ministry of Education and Science of the Russian Federation in the framework of MegaGrant no. 075-15-2019-1926.

We theoretically analyze this curvature considering two loss functions: *distortion*, which is widely used in embedding literature, and *threshold-based* loss (see Sect. 2), which is more suitable for some practical applications. We prove that these loss functions lead to fundamentally different graph curvatures. We also conduct a systematic analysis of several existing estimators of global curvature, in particular, the ones based on the well-known Ollivier-Ricci and Forman-Ricci *local* graph curvatures. We prove that all these notions give curvatures that are far from the optimal curvature for embedding the graph (in particular, because they are dimension-independent).

One insight, which we gained based on our theoretical analysis, is the fact that the optimal curvature is often smaller for smaller dimensions. In other words, while hyperbolic space may be needed for some small dimension, Euclidean space may be enough for a larger one.

Our analysis shows that the problem of estimating the global graph curvature is non-trivial. Moreover, it provides important first steps towards understanding this concept from both theoretical and practical aspects. This will aid future researchers studying this important topic.

2 Background and Related Work

Graph Embeddings. For an unweighted connected graph $G = (V, E)$ equipped with the shortest path distance metric d_G, a graph embedding f is a map $f : V \rightarrow U$, where U is a metric space. We refer to [2] for a survey of several graph embedding methods and their applications.

The goal of an embedding is to preserve some structural properties of a graph. Depending on the application, different loss/quality functions are used to measure the quality of a graph embedding. The most well-known is *distortion*:[1]

$$D(f) = \frac{1}{\binom{n}{2}} \sum_{u \neq v} \frac{|d(f(u), f(v)) - d_G(u, v)|}{d_G(u, v)},$$

where d denotes the distance metric in U.

Distortion is a global metric, it takes into account all graph distances. However, in some practical applications, it may not be the best choice. For example, in recommendation tasks, we usually deal with a partially observed graph, so a huge graph distance between nodes in the observed part does not necessarily mean that the nodes are not connected by a short path in the full graph. Additionally, as we shall see in Sect. 5.1, graph distances are hard to preserve: there are simple graphs on just 4 nodes that can be perfectly embedded only in a space of curvature $-\infty$ for any dimension.

Another measure, often used for embeddings, is Mean Average Precision (MAP), which, for a given node, compares the distance-based ranking of other embedded nodes with the graph-neighborhood-based ranking. We do not consider MAP in our analysis since it cares only about the order and so is curvature-invariant. Indeed, changing curvature is equivalent to changing scale, so for MAP

---

[1] There are other definitions of distortion in the literature, see, e.g., [17].

it is sufficient to consider only the curvatures -1, 0, 1, corresponding to hyperbolic, Euclidean and spherical spaces. Moreover, by considering a small enough region in hyperbolic or spherical space we get geometry similar to the Euclidean one, so for MAP it is important to distinguish only between -1 and 1.

We also consider the following class of *threshold-based* loss functions. Given an embedding f of a graph G, we (re)construct a graph G' in the following way: v and u are connected in G' iff $d(f(v), f(u)) \leq 1$. Then, any loss function which is based on the comparison of G and G' is called threshold-based. Such loss functions are natural in many applications (graph reconstruction, link prediction, recommendations). While many particular choices are possible, our theoretical results hold for *any* threshold-based loss function.

Hyperbolic and Spherical Spaces. For many years, Euclidean space was the primary choice for data embeddings [2]. However, it turned out that many observed datasets are well fitted into hyperbolic space [8]. In particular, hyperbolic embeddings can improve state-of-the-art quality in several practical tasks, e.g., lexical entailment and link prediction [12]. On the other hand, spherical spaces are also used for embeddings [9]. Going even further, [5] suggests mixed spaces: product manifolds combining multiple copies of spherical, hyperbolic, and Euclidean spaces.

The main advantage of hyperbolic space is that it is "larger": the volume of a ball grows exponentially with its radius. Hence, such spaces are well suited for tree-like structures. On the other hand, spherical spaces are suitable for embedding cycles [5]. Spherical and hyperbolic spaces are parametrized by *curvature c*, which is positive for spherical space and negative for hyperbolic space. As $c \to 0$, geometry of both these spaces becomes similar to the Euclidean one. We discuss some geometrical properties of these different spaces in Appendix A.

3   Local Graph Curvatures

While in this paper we analyze *global* graph curvature, there are several *local* ones proposed in the literature. Many of them are based on the notion of sectional curvature and Ricci curvature defined for Riemannian manifolds. Intuitively, Ricci curvature controls whether the distance between small spheres is smaller or larger than the distance between the centers of the spheres. For example, Ricci curvature is positive if small spheres are closer than their centers are. We refer to [6,15] for more details on Ricci curvature.

Ollivier Curvature. Ollivier curvature translates the definition of Ricci curvature to graphs. Again, the idea is to compare the distance between two small balls with the distance between their centers. The distance between balls is defined by the well-known optimal transport distance (a.k.a. Wasserstein distance or earthmover distance). Formally, for a graph G we consider the shortest path metric on G, denoted by d_G, and let W_1^G denote the Wasserstein metric with respect to the metric space (G, d_G). Furthermore, for each node v we let m_v denote the uniform probability measure on the neighbors of v, i.e., $m_v(u) = \frac{1_{u \sim v}}{\deg(v)}$, where $\deg(v)$ denotes the degree of v. Then, the classic definition[2] of Ollivier curvature between two neighboring nodes $v \sim u$ in G is defined as

$$\kappa_G(u, v) = 1 - W_1^G(m_v, m_u). \tag{1}$$

We note that Ollivier curvature always belongs to the interval $[-2, 1]$ [7].

Forman Curvature. Forman curvature [20] is based on the discretization of Ricci curvature proposed by [1]. It is defined for a general weighted graph G, with both node and edge weights. When the graph G is not weighted, the definition becomes:

$$F_G(u, v) = 4 - (\deg(v) + \deg(u)). \tag{2}$$

Forman was interested in a general discretization of curvature for Riemannian manifolds and his formula includes faces of any dimension. Although this can be translated to graphs [21], the formula becomes quite cumbersome. Therefore, in Eq. 2 only 1-dimensional faces (edges) are included. One can extend this expression by including higher dimensional faces. This was considered in [18], where 2-dimensional faces on three nodes (triangles) were included. In the case of an unweighted graph, we then obtain

$$\hat{F}_G(u, v) = F(u, v) + 3\Delta_{uv} = 4 - \deg(v) - \deg(u) + 3\Delta_{uv}, \tag{3}$$

where Δ_{uv} is the number of triangles that contain the edge (u, v).

Based on the definitions, both Forman curvatures, especially $F_G(u, v)$, are expected to often be highly negative (see Sect. 5).

Heuristic Sectional Curvature. A different notion of curvature used by [5] is based on the following geometric fact. Let abc be a geodesic triangle and let m be the (geodesic) midpoint of bc. Then the value $d(a, m)^2 + \frac{d(b,c)^2}{4} - \frac{d(a,b)^2 + d(a,c)^2}{2}$ is equal to zero in euclidean space, is positive in spherical space and negative in hyperbolic space.

For graphs, let v be a node in G, b, c neighbors of v and a any other node. Then, we define

$$\xi_G(v; b, c; a) = \frac{1}{2d_G(a, v)} \left(d_G(a, v)^2 + \frac{d_G(b, c)^2}{4} - \frac{d_G(a, b)^2 + d_G(a, c)^2}{2} \right). \tag{4}$$

This resembles the formula above with $m = v$ and the normalization constant $2d_G(v, a)$ is included to yield the right scalings for trees and cycles. To define the graph sectional curvature of a node v and its neighbors b, c, we average $\xi_G(v; b, c; a)$ over all possible a: $\xi_G(v; b, c) = \frac{1}{|V|-3} \sum_{a \in G \setminus \{v, b, c\}} \xi_G(v; b, c; a)$.[3]

---

[2] Note that Ollivier curvature is defined in much more generality in terms of metrics and random walks [14]. Thus, different version on graphs can be considered. Equation (1) corresponds to the classical choices of graph distance and random walk.

[3] We assume that a does not coincide with b or c, which does not affect the average much, but makes our results in Sect. 5 more succinct.

4   Global Graph Curvature

The problem with these different notions of *local* graph curvature is that they cannot be easily used in practical applications, where data is usually embedded in a space of *constant* curvature. Hence, a *global* notion of curvature is needed. In this section, we propose a practice-driven definition of global graph curvature and discuss how to estimate this curvature based on local notions.

4.1   Definition

For a graph G, let $f(G)$ be an embedding of this graph into a d-dimensional space of constant curvature c (spherical, Euclidean or hyperbolic). Assume that we are given a loss function $L(f)$ for the embedding task (see Sect. 2). Now, let $L_{opt}(G, d, c)$ be the optimal loss for given d and c: $L_{opt}(G, d, c) = \min_f L(f)$. Then, we define d-dimensional curvature of G in the following way:

$$C_d^L(G) = \arg\min_c L_{opt}(G, d, c). \tag{5}$$

Note that there may be several values of curvature c delivering the minimum of $L_{opt}(G, d, c)$, in this case we say that $C_d^L(G)$ consists of all such points.[4]

Below we analyze global curvatures based on distortion ($C_d^{dist}(G)$) and threshold-based ($C_d^{thr}(G)$) loss functions. In the latter case, our theoretical results apply to any threshold-based loss, since $L_{opt}(G, d, c)$ reaches its minimum on "perfect" embeddings, where we precisely reconstruct the graph G.

4.2   Approximations

Let us discuss how local graph curvatures can be used to estimate the global one. In all cases, the standard practice is to average edge or sectional curvature over the graph.

Ollivier Curvature. $\kappa(G) = \frac{1}{|E|} \sum\limits_{u \sim v} \kappa_G(u, v)$.

Forman Curvature. $F(G) = \frac{1}{|E|} \sum\limits_{u \sim v} F_G(u, v)$, $\hat{F}(G) = \frac{1}{|E|} \sum\limits_{u \sim v} \hat{F}_G(u, v)$.

Average Sectional Curvature. Let P_3 denote the number of paths of length 3 in G, then $\xi(G) = \frac{1}{P_3} \sum\limits_{v \in V} \sum\limits_{b < c : b, c \in N(v)} \xi_G(v; b, c)$.

It is important to note that all curvatures discussed above do *not* depend on dimension d and loss function L. However, as we show below, global curvature defined in Sect. 4.1 significantly depends on them.

---

[4] Further we slightly abuse notation by writing that $C_d^L(G)$ is a real value if such c is unique and a set of values otherwise.

There is also a concept of Gromov's hyperbolicity [3], which is sometimes used to decide whether it is reasonable to embed a graph to a hyperbolic space. A metric has δ-hyperbolicity if all geodesic triangles are δ-slim: for any three points a, b, c, the shortest paths between them satisfy the following property: any point on one path is within distance δ from the closest point on the other two paths. However, firstly, such estimator cannot be easily converted to curvature, and, secondly, it does not say anything about the embedding in spherical spaces [11].

5   Theoretical Analysis of Global Curvature

To better understand the performance of the proposed approximations of global graph curvature, we consider several basic graphs and compare their global curvature and approximations. By studying these graphs we also gain insights into how classic graph topologies influence the curvature of the space in which they can be properly embedded.

5.1   Star S_n

It is pointed out in numerous papers that trees are negatively curved. We analyze this theoretically and start with the simplest tree: one central node and n leaves. We denote this graph by S_n and assume that $n \geq 3$.

Ollivier Curvature. Consider any tree graph T, let v, u be two neighbors. Then Proposition 2 in [7] states that

$$\kappa_T(u, v) = -2 \left(1 - \frac{1}{\deg(v)} - \frac{1}{\deg(u)} \right)^+ , \tag{6}$$

where $t^+ = \max\{0, t\}$. In particular, if either $\deg(v) = 1$ or $\deg(u) = 1$, then $\kappa_T(u, v) = 0$. As a result, for a star we have $\kappa_{S_n}(u, v) = 0$, so $\kappa(S_n) = 0$ and stars are *not* negatively curved according to Ollivier curvature.

Forman Curvature. If follows from (2) and (3) that $F(S_n) = \hat{F}(S_n) = 3 - n$, so stars are highly negatively curved for large n according to Forman curvature.

Average Sectional Curvature. Heuristic sectional curvature is defined for a node and its two neighbors. In case of a star we can only take a central node v and two neighboring ones b and c. For any other node a we obtain $\xi_{S_n}(v; b, c; a) = -1$. Therefore, by averaging we obtain $\xi(S_n) = -1$.

Distortion-Based Curvature. The following theorem holds.

Theorem 1. *Let d and n be fixed. If c is bounded below by a constant, then $D(S_n)$ is also bounded below by a constant. If $c \to -\infty$, then the optimal distortion $D_{opt}(S_n) = \Theta\left(\frac{1}{\sqrt{-c}}\right)$. Therefore, for any S_n we have $C_d^{dist}(S_n) = -\infty$.*

The intuition behind this result is the following: we cannot embed a star S_3 with zero distortion into any space of constant curvature and any dimension, because in case of zero distortion the central node v has to lie on the geodesics between all pairs of leaves, so all 4 nodes have to belong to one geodesics, which is impossible. Moreover, the same problem occurs if any graph G contains S_3 as an induced subgraph. On the other hand, if $c \to -\infty$, we can spread all leaves of S_n uniformly on a circle of radius 1 around the central node and distortion of such construction will tend to zero since the distance between the pairs of leaves will tend to 2 (triangles become thinner). The formal proof of Theorem 1 can be found in Appendix B. Further we will see that if we minimize a threshold-based loss, then any tree can be perfectly embedded with $d = 2$.

Threshold-Based Curvature. Here we have the following theorem.

Theorem 2. $C_d^{thr}(S_n) = (-\infty, C)$ *for some $C = C(n, d)$, which increases with d and decreases with n. In particular, for $d = 2$, if $n < 6$, then $C = \left(\arccos\frac{\cos\frac{2\pi}{n}}{1 - \cos\frac{2\pi}{n}}\right)^2$; if $n = 6$, then $C = 0$; if $n > 6$, then $C = -\left(2\arccosh\frac{1}{2\sin\frac{\pi}{n}}\right)^2$.*

Proof. We show that for any n and d and some curvature c there exists a perfect embedding (preserving all edges). Therefore $C_d^{dist}(S_n)$ consists of curvatures for which such perfect embedding exists. Note that if there exists a perfect embedding f for some curvature c, then there exists a perfect embedding for any curvature $c' < c$. Indeed, w.l.o.g., we assume that the central node v is mapped to the origin of a hyperspherical coordinate system and other points $v_1, \ldots, v_n$ can be described by their radii and angles. We know that the distance between v and any v_i is at most 1 and the distance between any pair v_i, v_j is larger than one. Now we change curvature to $c' < c$ and keep hyperspherical coordinates the same. Then the distance between v and any v_i does not change, while the distance between nodes v_i, v_j increases.

It is easy to see that C increases with d: if there exists an embedding to some dimension d, then, obviously, the same embedding works for $d' > d$. Further, C decreases with n since if there exists an embedding of S_n, then we can easily construct an embedding of $S_{n'}$ for $n' < n$ by removing some nodes.

The rest of the proof (constructing a perfect embedding for $d = 2$) is technical and can be found in Appendix C. The main idea is that when we embed a star aiming to minimize a threshold-based loss, we just need a curvature to be small enough to spread all neighbors of a central node sufficiently far away from each other.

5.2   Tree T_b with Branching Factor b

We consider a tree T_b, $b \geq 2$. For symmetry, assume that the first node has $b+1$ children, while all other nodes have a parent and b children. For Ollivier and Forman curvatures, we give the results for the case when the depth M of T_b tends to infinity. For threshold-based curvature we will consider a tree of infinite depth. In the other cases, our statements hold for any finite tree T.

Ollivier Curvature. Let $E_b(M)$ denote the number of edges in the tree $T_b(M)$ of depth $M \geq 2$ and observe that $E_b(M) = (b+1)\sum_{m=0}^{M-1} b^m$. There are two types of edges: those adjacent to a leaf node and those not adjacent to a leaf node. For the former, we have that one node has degree 1 and hence (6) implies that Ollivier curvature of these edges is 0. For edges (u, v) that are not adjacent to a leaf node (6) implies that $\kappa_{T_b(M)}(u, v) = -2\left(1 - \frac{1}{b+1} - \frac{1}{b+1}\right) = -\frac{2(b-1)}{b+1}$. Since there are $E_b(M-1)$ edges of this type we get

$$\kappa(T_b(M)) = -\frac{2(b-1)}{b+1}\frac{E_b(M-1)}{E_b(M)},$$

which is negative for all $M \geq 2$.

Finally we note that $E_b(M-1) = \frac{E_b(M)-(b+1)}{b}$ and $E_b(M) \to \infty$ as $M \to \infty$, so that

$$\kappa(T_b(M)) = -\frac{2(b-1)}{b+1}\left(\frac{1}{b} - \frac{b+1}{bE_b(M)}\right) \to -\frac{2(b-1)}{b(b+1)}.$$

Forman Curvature. It is easy to see that for an edge (u, v) adjacent to a leaf node $F_{T_b(M)}(u, v) = \hat{F}_{T_b(M)}(u, v) = 2 - b$ while edges (u', v') that are not adjacent to a leaf node $F_{T_b(M)}(u, v) = \hat{F}_{T_b(M)}(u, v) = 2 - 2b$. Hence we obtain that

$$F(T_b(M)) = \hat{F}(T_b(M))$$

$$= \frac{1}{E_b(M)}\left(E_b(M-1)(2-2b) + (E_b(M) - E_b(M-1))(2-b)\right)$$

$$= \frac{1}{E_b(M)}\left(E_b(M)(2-b) - E_b(M-1)b\right) = (1-b) + \frac{b-1}{E_b(M)},$$

where we used the relation between $E_b(M-1)$ and $E_b(M)$. From this it follows that

$$F(T_b(M)) = \hat{F}(T_b(M)) \to (1-b),$$

which is negative.

Average Sectional Curvature. In contrast to Ollivier and Forman curvatures, heuristic sectional curvature is global, i.e., it depends on the whole graph, which has to be finite. Note that for any tree, to compute sectional curvature, we average 0 and -1. As a result, for any tree T we have $\xi(T) \in [-1, 0]$ [5].

Distortion-Based Curvature. On the one hand, our result for S_n implies that if a graph contains S_3, then it cannot be embedded with zero distortion in any space. One the other hand, [19] proves that if we scale all edges by a sufficiently large factor τ, then the obtained tree can be embedded in the hyperbolic plane with distortion at most $1 + \varepsilon$ with arbitrary small ε. Note that multiplying graph edges by τ is equivalent to changing curvature from 1 to τ^2. As a result, [19] proves that we can achieve an arbitrary small distortion if $c \to -\infty$. Hence, $C_d^{dist}(T) = -\infty$ for any T.

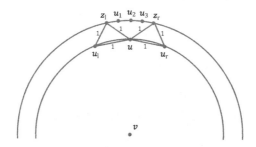

Fig. 1. Threshold-based embedding of trees. The nodes u_l, u and u_r are at level l while u_1, u_2, u_3 are at level $l + 1$. This figure illustrates an intuitive idea of the proof construction, the exact shape of the circles and triangles will depend on the specific model of hyperbolic space used.

Threshold-Based Curvature

Theorem 3. $C_d^{thr}(T_b) = (-\infty, C)$ for some $C = C(b, d)$, which increases with d and decreases with b. In particular, $C(b, 2) \geq - \left(\dfrac{2 \log b}{2 \operatorname{arccosh} \frac{\cosh 1}{\cosh 1/2} - 1} \right)^2$.

Actually, the bound above holds for any tree whose branching is bounded by b. Interestingly, while it is often claimed that trees are intrinsically hyperbolic [8,12], to the best of our knowledge, we are the first to formally prove that trees can be perfectly embedded in a hyperbolic plane of some curvature.

Proof. First, recall that Appendix A lists some geometric properties of hyperbolic space which we use throughout this proof.

As for S_n, we prove that for T_b a perfect embedding exists for $d = 2$ and some curvature c. Then, similarly to the previous section, it is clear that C increases with b and decreases with n.

For the lower bound on C, we have to guarantee that an embedding exists. For this, we provide the following construction (see Fig. 1 for an illustration). (Below we assume that the curvature is large enough for our construction to work, then we estimate the required curvature.) First, we take the node v and consider a circle of radius 1 around this node. We spread $b+1$ neighbors uniformly around this node. For our construction to work, we need all distances between these nodes to be larger than 1. Now, at some step of the algorithm, assume that we have all nodes at level l placed at some circle centered at v and all distances between the nodes at level l are larger than 1. Our aim is then to find positions for all nodes at level $l + 1$.

Let us take any node at l-th level. Consider two points u_l and u_r on the same circle at distance 1 from the node u to the left and to the right, respectively. Let u, u_l, z_l and u, u_r, z_r form equilateral triangles (with sides equal to 1). Then we let the points at $l + 1$-th level to be spread on the circle centred at v and passing through z_l and z_r. The children of u $(u_1, \ldots, u_b)$ will be placed on the circular arc between z_l and z_r. As usual, we want u_i and u_{i+1} to be at distance at least 1 from each other. Moreover, they have to be at distance at least 1 from children of other nodes. Also, note that placing $u_1, \ldots, u_b$ between z_l and z_r guarantees that these nodes are closer than 1 to their parent node u but at the same time at a distance larger than 1 from all other nodes at l-th level. Also, all points at l-th level are far enough from points at $l + 2$-th level.

It remains to find a maximum curvature such that the required conditions are satisfied. Let r and r' be radii of circles at l-th and $l + 1$-th levels and let $2\alpha = \angle u_l v u$. We know (the law of cosines and $\cos 2\alpha = 1 - 2\sin^2 \alpha$) that

$$\cosh \frac{1}{R} = 1 + 2\sinh\left(\frac{r}{R}\right) \sin^2 \alpha \, , \tag{7}$$

where $R = 1/\sqrt{-C}$ (see Appendix A). So, the only condition we need for the whole procedure to work is that we have enough space on the circular arc for placing b nodes there:

$$\cosh \frac{1}{R} \leq 1 + 2\sinh\left(\frac{r'}{R}\right) \sin^2 \frac{\alpha}{b} \, .$$

We note that $\sin^2 \frac{\alpha}{b} \geq \frac{\sin^2 \alpha}{b^2}$ for all $b \geq 1$. Therefore, it is sufficient to have

$$\cosh \frac{1}{R} \leq 1 + 2\sinh\left(\frac{r'}{R}\right) \cdot \frac{\sin^2 \alpha}{b^2} \, . \tag{8}$$

Combining (7) and (8), we obtain:

$$\sinh \frac{r'}{R} \geq b^2 \cdot \sinh \frac{r}{R}.$$

To achieve this, it is sufficient to have:

$$\frac{r' - r}{R} \geq 2\log b,$$

$$R \leq \frac{r' - r}{2\log b}.$$

It remains to find the lower bound for $r' - r$ and it is easy to see that $r' - r$ decreases with r. Therefore, it is sufficient to consider only the second step of the construction procedure, when we move from the circle of radius 1 to the next one. In this case, $r = 1$ and

$$r' = 2\operatorname{arccosh}\left(\frac{\cosh 1}{\cosh 0.5}\right).$$

So, we have

$$-C = \frac{1}{R^2} \le \left(\frac{2\log b}{2\operatorname{arccosh}\left(\frac{\cosh 1}{\cosh 0.5}\right) - 1}\right)^2.$$

5.3   Complete Graph K_n

Ollivier Curvature. For any two nodes u and v, it follows from Example 1 in [7] that $\kappa_{K_n}(v, u) = \frac{n-2}{n-1}$. Thus, $\kappa(K_n) = \frac{n-2}{n-1}$ and it tends to 1 as $n \to \infty$.

Forman Curvature. Simple computations yield: $F(K_n) = 6 - 2n$, $\hat{F}(K_n) = n$, i.e., we get either highly positive or highly negative value.   ◂

Average Sectional Curvature. It is easy to compute that $\xi(K_n) = \frac{1}{8}$.

Distortion-Based Curvature. The following theorem analyzes $C_d^{dist}(K_n)$ if $d = n - 2$.

Theorem 4. $C_{n-2}^{dist}(K_n) = \left\{ -\infty, 4\left(\arcsin\sqrt{\frac{n}{2(n-1)}}\right)^2\right\}.$

Proof. If $d = n - 2$, then we are given a $(n-1)$-simplex, which can be embedded into $n-2$-dimensional spherical space. Indeed, the radius of circumscribed hypersphere for the $(n-1)$-simplex with side length a is known to be $R = a\sqrt{\frac{n-1}{2n}}$. Since we want the *spherical* distance between all points to be equal to one, we need to choose a accordingly:

$$\sin\frac{\alpha}{2} = \frac{a}{2R}\quad\text{for}\quad \alpha = \frac{1}{R}.$$

Here α corresponds to the angle giving the arc length 1, while the condition on $\sin\frac{\alpha}{2}$ relates α and a since α is the angle in a triangle with side lengths α, R, R. This implies that

$$2R\arcsin\frac{a}{2R} = 1,$$

and solving this equation for a yields

$$a = \sqrt{\frac{2n}{(n-1)}}\frac{1}{2\arcsin\sqrt{\frac{n}{2(n-1)}}}.$$

Plugging this back into the formula for the radius R we obtain

$$R = \frac{1}{2 \arcsin \sqrt{\frac{n}{2(n-1)}}},$$

so that we have

$$C_{n-2}^{dist}(K_n) = 4 \left(\arcsin \sqrt{\frac{n}{2(n-1)}} \right)^2.$$

Finally, let us show that if $c \to -\infty$, then optimal distortion $D_{opt}(K_n) \to 0$. This result follows from the fact that $D_{opt}(S_n) \to 0$, because to embed a clique, it is sufficient to embed a star on $n+1$ node with edge lengths $1/2$ and then remove the central node.

Threshold-Based Curvature. $C_d^{thr}(K_n) = \mathbb{R}$, since we can embed any complete graph perfectly by mapping all nodes to one point.

5.4   Cycle Graph C_n

We consider a cycle C_n with $n \geq 4$.

Ollivier Curvature. Let $v \sim u$ be two neighbors. Then it is easy to see that $W_1^G(m_u, m_v) = 1$ and hence $\kappa_G(u, v) = 0$. Thus, $\kappa(C_n) = 0$.

Forman Curvature. Similarly, it is easy to see that $F(C_n) = \hat{F}(C_n) = 0$.

Average Sectional Curvature. If n is even, then $\xi_{C_n}(v; b, c; a) = 0$ for all points except the one diametrically opposite to v for which we have $\xi_{C_n}(v; b, c; a) = 1$. If n is odd, then for two points we have $\xi_{C_n}(v; b, c; a) = \frac{n}{2(n-1)}$. As a result, $\xi(C_n) = \frac{1}{n-3}$ for even n and $\xi(C_n) = \frac{n}{(n-1)(n-3)}$ for odd n.

Distortion-Based Curvature. Here we have that $C_d^{dist}(C_n) = \left(\frac{2\pi}{n} \right)^2$. Indeed, if we consider any three consequent nodes, then the middle one should lie on the geodetic between the other two. So, they all lie on a great circle (of length n) from which the result follows.

Threshold-Based Curvature. It is easy to see that $C_d^{thr}(C_n) = (-\infty, C)$ with some $C > 0$, which decreases with n and increases with d. A simple lower bound for C is $C \geq \left(\frac{4\pi}{n} \right)^2$, since for such curvature we can embed all nodes into a great circle with distances $1/2$ between the closest ones.

5.5   Complete Bipartite Graph $K_{l,m}$

W.l.o.g. we assume that $l \geq m \geq 2$ (the remaining cases are stars and are already considered).

Ollivier Curvature. We prove the following lemma.

Lemma 1. $\kappa(K_{l,m}) = 0$.

Proof. Denote the node sets in $K_{l,m}$ by $U := \{u_1, \ldots, u_l\}$ and $V := \{v_1, \ldots, v_m\}$. We will prove that for any edge (u,v), $W_1^{K_{l,m}}(m_u, m_v) = 1$, which then implies that $\kappa(K_{l,m}) = 0$. For this we use the dual representation for the Wasserstein distance, see [14] for more details. This states that on the one hand

$$W_1^G(m_u, m_v) = \inf_{\rho} \sum_{v' \sim v} \sum_{u' \sim u} d_G(v', u') \rho(v', u'),$$

where the infimum is taken over all joint probability measures on the product of the neighborhoods of v and u, while on the other hand

$$W_1^G(m_u, m_v) = \sup_f \left(\frac{1}{\deg(v)} \sum_{v' \sim v} f(v') - \frac{1}{\deg(u)} \sum_{u' \sim u} f(u') \right),$$

where the supremum is taken over all 1-Lipschitz functions, i.e., $|f(u) - f(v)| \leq d_G(u,v)$.

Note that for any $u \in U, v \in V$ the joint neighborhood is $V \times U$. First we establish an upper bound by considering the product joint probability density on $V \times U$ $\rho(x,y) = \frac{1}{ml}$. It then follows that

$$W_1^{K_{l,m}}(m_u, m_v) \leq \sum_{i=1}^{m} \sum_{j=1}^{l} d_G(v_i, u_j) \rho(v_i, u_j) = 1.$$

For the lower bound, we define the function

$$f(z) = \begin{cases} 2 & \text{if } z \in V, \\ 1 & \text{if } z \in U. \end{cases}$$

Observe that if $u \in U$ and $v \in V$ then $|f(u) - f(v)| = 1 = d_G(u,v)$. On the other hand, if $u, u' \in U$ then $|f(u) - f(u')| = 0 \leq 2 = d_G(u, u')$ and similar for $v, v' \in V$. Thus we conclude that f is 1-Lipschitz. It now follows that

$$W_1^{K_{l,m}}(m_u, m_v) \geq \frac{1}{m} \sum_{i=1}^{m} f(v_i) - \frac{1}{l} \sum_{j=1}^{l} f(u_j) = 1,$$

which completes the proof.

Forman Curvature. It is easy to see that $F(K_{l,m}) = \hat{F}(K_{l,m}) = 4 - l - m$.

Average Sectional Curvature. The following lemma holds.

Lemma 2. $\xi(K_{l,m}) = \frac{-(l-m)^2 + m + l - 2}{(m+l-2)(l+m-3)}$. *In particular, if* $m = l$ *we get* $\xi(K_{l,m}) = \frac{1}{2m-3}$.

This lemma implies that for balanced complete bipartite graphs $\xi(K_{l,m})$ is positive, but tends to zero as the graph grows.

Proof. If v and a are in the same part of the bipartite graph, then $\xi_{K_{l,m}}(v; b, c; a) = 1$, otherwise $\xi_{K_{l,m}}(v; b, c; a) = -1$. Therefore, if v belongs to the part of size l, sectional curvature is $\xi_{K_{l,m}}(v; b, c) = \frac{l-m+1}{l+m-3}$, otherwise it is $\xi_{K_{l,m}}(v; b, c) = \frac{m-l+1}{l+m-3}$. As a result, by averaging over all triplets, we get

$$\xi(K_{l,m}) = \frac{1}{l\binom{m}{2} + m\binom{l}{2}} \left(l\binom{m}{2} \frac{l-m+1}{l+m-3} + m\binom{l}{2} \frac{m-l+1}{l+m-3} \right)$$

$$= \frac{-(l-m)^2 + m + l - 2}{(m+l-2)(l+m-3)}.$$

Distortion-Based Curvature. We prove the following simple proposition.

Proposition 1. *For any* d, $C_d^{dist}(K_{2,2}) = \left(\frac{\pi}{2}\right)^2 \approx 2.47$ *and* $K_{2,2}$ *is the only complete bipartite graph (with at least two nodes in each part) for which zero distortion is achievable.*

Proof. Indeed, the result for $K_{2,2}$ follows from the corresponding result on cycle C_4. Moreover, if for $K_{l,m}$ we have $l \geq 3$ and $m \geq 2$, then for any two nodes in the part of size l there are at least 2 different geodesics of length 2 between them. Therefore, all such pairs lie at opposite poles of the hypersphere, which is impossible since $l \geq 3$.

6   Conclusion

We introduced a concept of global graph curvature motivated by the practical task of embedding graphs. This curvature depends on the loss function and space dimension. To get an intuition about how global graph curvature behaves, we theoretically analyzed it for several simple graphs. We compared the global graph curvature and several approximations based on well-known local graph curvatures and showed that they essentially differ. We demonstrated that dimensionality and the choice of a loss function fundamentally affect the global curvature and, in particular, when dimension is larger the optimal curvature usually becomes less negative. Our work shows that the problem of finding the right space for graph embedding is interesting and non-trivial and we hope our results will encourage further research on global graph curvature.

A   Geometrical Properties of Spaces of Constant Curvature

In this section, we recall some useful equalities which will be used throughout the proofs.

We use notation R, where $R = \frac{1}{\sqrt{c}}$ in spherical space (corresponds to the radius of a sphere) and $R = \frac{1}{\sqrt{-c}}$ in the hyperbolic case (can be considered as a scaling factor compared to the space of curvature -1).

Law of Cosines. Let us consider a triangle with angles A, B, C and the lengths of opposite sides a, b, c, respectively.

In Euclidean, space we have:

$$c^2 = a^2 + b^2 - 2\,a\,b\cos C\,.$$

In spherical space, the first law of cosines is:

$$\cos\frac{c}{R} = \cos\frac{a}{R}\cos\frac{b}{R} + \sin\frac{a}{R}\sin\frac{b}{R}\cos C\,,$$

and the second law of cosines is:

$$\cos C = -\cos A\cos B + \sin A\sin B\cos\frac{c}{R}\,.$$

In hyperbolic space, we have

$$\cosh\frac{c}{R} = \cosh\frac{a}{R}\cosh\frac{b}{R} - \sinh\frac{a}{R}\sinh\frac{b}{R}\cos C\,.$$

Equilateral Triangle. The following equalities follow from the corresponding laws of cosines, assuming that all sides (and angles) are equal.

For hyperbolic space:

$$\cosh\frac{a}{2R} = \frac{1}{2\sin\frac{A}{2}}\,. \tag{9}$$

For spherical space:

$$\cos\frac{a}{R} = \frac{\cos A}{1 - \cos A}\,. \tag{10}$$

Area and Volume of Hypersphere. Let $S_d(r)$ and $V_d(r)$ denote area of a hypersphere and volume of a ball of radius r in d-dimensional space.

In euclidean space,

$$S_d(r) = d\,C_d\,r^{d-1}\,,$$

$$V_d(r) = C_d\,r^d\,,$$

where

$$C_d = \frac{\pi^{d/2}}{\Gamma(\frac{d}{2}+1)}\,.$$

In spherical space, sphere of radius r is isometric to Euclidean sphere of radius $R \sin \frac{r}{R}$. Therefore, the area is

$$S_d(r) = d\, C_d \left(R \sin \frac{r}{R} \right)^{d-1},$$

$$V_d(r) = d\, C_d\, R^d \int_0^r \left(\sin \frac{x}{R} \right)^{d-1} dx.$$

Similarly, in hyperbolic space,

$$S_d(r) = d\, C_d \left(R \sinh \frac{r}{R} \right)^{d-1},$$

$$V_d(r) = d\, C_d\, R^d \int_0^r \left(\sinh \frac{x}{R} \right)^{d-1} dx.$$

B   Proof of Theorem 1

First, let us analyze the lower bound on distortion. Recall that distortion of a graph is the average distortion over all pairs of nodes. Let v be the central node and $v_1, \ldots, v_n$ $(n \geq 3)$ be its neighbors. For any embedding f, we have

$$D(S_n) = \frac{1}{\binom{n+1}{2}} \left(\sum_{v_i} |d(f(v), f(v_i)) - 1| + \sum_{v_i \neq v_j} \frac{|d(f(v_i), f(v_j)) - 2|}{2} \right)$$

$$= \frac{1}{\binom{n+1}{2}} \sum_{1 \leq i_1 < i_2 < i_3 \leq n} \left(\sum_{1 \leq j \leq 3} \frac{|d(f(v_{i_j}), f(v)) - 1|}{\binom{n-1}{2}} + \sum_{1 \leq j < k \leq 3} \frac{|d(f(v_{i_j}), f(v_{i_k})) - 2|}{2(n-2)} \right).$$

Let D_{min} be the minimum value of the following weighted distortion of a star with 3 leaves:

$$D_{min} = \min_f \sum_{1 \leq j \leq 3} \frac{|d(f(v_j), f(v)) - 1|}{(n-1)/4} + \sum_{1 \leq j < k \leq 3} |d(f(v_j), f(v_k)) - 2|,$$

then

$$D(S_n) \geq \frac{\binom{n}{3}}{2(n-2)\binom{n+1}{2}} D_{min} = \frac{(n-1)D_{min}}{6(n+1)}. \tag{11}$$

Hence, it remains to find a lower bound on D_{min}, i.e., a lower bound for a weighted distortion of S_3 with central node v and three leaves v_1, v_2, v_3. If we consider three angles at the node v, then at least one of them is $\alpha \leq 2\pi/3$, so we can get a lower bound by only considering this triangle, which is, w.l.o.g., formed by v, v_1, v_2.

$$D_{min} \geq |d(f(v_1), f(v_2)) - 2| + \frac{|d(f(v_1), f(v)) - 1| + |d(f(v_2), f(v)) - 1|}{(n-1)/4}.$$

Denote $d(f(v_1), f(v)) = x = 1 + \varepsilon$, $d(f(v_2), f(v)) = y = 1 + \delta$, $d(f(v_1), f(v_2)) = z = 2 + \varepsilon + \delta - \varphi$ with some ε, δ and some $\varphi > 0$ (from triangle inequality). Assume that $|\varepsilon| < 1/2$ and $|\delta| < 1/2$ (otherwise the lower bound is trivial). Now we use the law of cosines to get a lower bound on φ. We consider Euclidean and hyperbolic spaces separately and note that the bound obtained in Euclidean space also holds in spherical spaces (with any c).

In Euclidean space, using triangle inequality, we get $\varphi = x + y - z > 0$. So, in Euclidean and spherical spaces φ is bounded below by a constant.

In hyperbolic space the law of cosines gives (we denote $R = \frac{1}{\sqrt{-c}}$):

$$\cosh \frac{z}{R} = \cosh \frac{x}{R} \cosh \frac{y}{R} - \sinh \frac{x}{R} \sinh \frac{y}{R} \cos \alpha,$$

from which, using $\cosh(x + y) = \cosh x \cosh y + \sinh x \sinh y$, we get

$$\cosh \frac{x + y}{R} - \cosh \frac{z}{R} = \sinh \frac{x}{R} \sinh \frac{y}{R}(1 + \cos \alpha).$$

If $R \to \infty$, then, similarly to Euclidean case, we get $\varphi = x + y - z = \Omega(1)$.

On the other hand, if $R \to 0$, we get $\varphi = \Omega(R)$. Note that $D_{min} \geq |z - 2| + \frac{|x-1|+|y-1|}{(n-1)/2} = |\varepsilon + \delta - \varphi| + \frac{|\varepsilon|+|\delta|}{(n-1)/2}$. This gives us a lower bound $D_{min} = \Omega(1/n)$ in spherical and Euclidean spaces and $D_{min} = \Omega(\min(R, 1)/n) = \Omega\left(\frac{1}{n \max(\sqrt{-c}, 1)}\right)$ in hyperbolic space. From this and (11) the bound on $D(S_n)$ follows.

Now, let us get an upper bound on optimal distortion $D_{opt}(S_n)$. To do this, we explicitly construct an embedding with sufficiently low distortion $D(S_n)$.

Let v be the central node, then we spread all other nodes uniformly on a 2-dimensional circle of radius 1 centred at v. The smallest angle between two points is $2\pi/n$. Therefore, from the law of cosines, the distance between leaves is at least k with

$$\cosh \frac{k}{R} = 1 + \left(1 - \cos \frac{2\pi}{n}\right) \sinh^2 \frac{1}{R}.$$

Note that for any two leaves v_i and v_j we have that $d(f(v_i), f(v_j)) \leq 2$. In particular, the closer two leaves are, the greater the difference $2 - d(f(v_i), f(v_j))$ is. Hence, the distance between adjacent leaves is the worst case and thus $D_{opt}(S_n)$ can be upper bounded as

$$D_{opt}(S_n) \leq \frac{\binom{n}{2}}{2\binom{n+1}{2}} \left(2 - R \cdot \operatorname{arccosh}\left(\left(1 - \cos \frac{2\pi}{n}\right) \sinh^2 \frac{1}{R} + 1\right)\right)$$

$$= \frac{(n-1)}{2(n+1)} \left(2 - R \cdot \operatorname{arccosh}\left(\left(1 - \cos \frac{2\pi}{n}\right) \sinh^2 \frac{1}{R} + 1\right)\right).$$

Note that $1 - \cos \frac{2\pi}{n} = \Theta\left(\frac{1}{n}\right)$ and $\sinh^2\left(\frac{1}{R}\right) = \Theta\left(e^{2/R}\right)$. Then, the value $\operatorname{arccosh}\left(\Theta\left(\frac{1}{n}e^{2/R}\right) + 1\right)$ behaves as $\sqrt{2e^{2/R}/n}$ if $2/R \ll \log n$ and as $\frac{2}{R} - \log n$

if $2/R \gg \log n$. Therefore, we get

$$D_{opt}(S_n) = O\left(R \log n\right) = O\left(\frac{\log n}{\sqrt{-C}}\right).$$

C   Proof of Theorem 2

Let us construct a perfect embedding of S_n for $d = 2$ and estimate the required curvature. Recall that in a perfect embedding all leaves have to be inside the ball of radius 1 around the central node v and also the distance between any two leaves has to be larger than one. It is easy to see that if we managed to spread n points inside the ball of radius 1 with distances more than 1 between them, then we can move each point along the radius up to distance 1 from v preserving this property. Therefore, it is sufficient to spread all points on a hypersphere.

First, assume that $n > 6$. In this case we have to consider only hyperbolic space, since n neighbors would not fit to a circle of radius 1 in neither spherical nor Euclidean spaces. We will find the largest curvature C which allows to have distance exactly 1 between the closest leaves. In this case we cannot embed S_n in a space of curvature C, but can embed in a space of any smaller curvature. We use (9) (where $R = \frac{1}{\sqrt{-C}}$) and let $\alpha = \frac{2\pi}{n}$:

$$\cosh\frac{1}{2R} = \frac{1}{2\sin\frac{\alpha}{2}},$$

$$R = \frac{1}{2\operatorname{arccosh}\frac{1}{2\sin\frac{\alpha}{2}}},$$

$$C = -\left(2\operatorname{arccosh}\frac{1}{2\sin\frac{\alpha}{2}}\right)^2.$$

Note that if n is large, then $\sin\frac{\alpha}{2} = \sin\frac{\pi}{n-1} \sim \frac{\pi}{n-1}$. Then, $\operatorname{arccosh}\frac{1}{2\sin\frac{\alpha}{2}} \sim \operatorname{arccosh}\frac{n-1}{2\pi} \sim \log n$, so we get $C \sim -4\log^2 n$.

Now, let us consider $n \leq 6$. Obviously, for $n = 6$ we have $C = 0$.

If $n < 6$, then $C > 0$. In this case we consider a spherical space and use the corresponding law of cosines (10):

$$\cos\frac{1}{R} = \frac{\cos\alpha}{1 - \cos\alpha},$$

$$C = \left(\arccos\frac{\cos\alpha}{1 - \cos\alpha}\right)^2.$$

References

1. Forman, R.: Bochner's method for cell complexes and combinatorial Ricci curvature. Discrete Comput. Geom. **29**(3), 323–374 (2003). https://doi.org/10.1007/s00454-002-0743-x
2. Goyal, P., Ferrara, E.: Graph embedding techniques, applications, and performance: a survey. Knowl.-Based Syst. **151**, 78–94 (2018). https://doi.org/10.1016/j.knosys.2018.03.022
3. Gromov, M.: Hyperbolic groups. In: Gersten, S.M. (ed.) Essays in Group Theory, vol. 8, pp. 75–263. Springer, New York (1987). https://doi.org/10.1007/978-1-4613-9586-7_3
4. Grover, A., Leskovec, J.: node2vec: scalable feature learning for networks. In: Proceedings of the 22nd ACM SIGKDD International Conference on Knowledge Discovery and Data Mining, pp. 855–864. ACM (2016)
5. Gu, A., Sala, F., Gunel, B., Ré, C.: Learning mixed-curvature representations in product spaces. In: ICLR (2019). https://openreview.net/pdf?id=HJxeWnCcF7
6. Jost, J.: Geometry and Physics. Springer, Heidelberg (2009). https://doi.org/10.1007/978-3-642-00541-1
7. Jost, J., Liu, S.: Ollivier's Ricci curvature, local clustering and curvature-dimension inequalities on graphs. Discrete Comput. Geom. **51**(2), 300–322 (2013). https://doi.org/10.1007/s00454-013-9558-1
8. Krioukov, D., Papadopoulos, F., Kitsak, M., Vahdat, A., Boguná, M.: Hyperbolic geometry of complex networks. Phys. Rev. E **82**(3), 036106 (2010)
9. Liu, W., Wen, Y., Yu, Z., Li, M., Raj, B., Song, L.: SphereFace: deep hypersphere embedding for face recognition. In: Proceedings of the IEEE Conference on Computer Vision and Pattern Recognition, pp. 212–220 (2017)
10. Mikolov, T., Sutskever, I., Chen, K., Corrado, G.S., Dean, J.: Distributed representations of words and phrases and their compositionality. In: Advances in Neural Information Processing Systems, pp. 3111–3119 (2013)
11. Ni, C.C., Lin, Y.Y., Gao, J., Gu, X.D., Saucan, E.: Ricci curvature of the internet topology. In: 2015 IEEE Conference on Computer Communications (INFOCOM), pp. 2758–2766. IEEE (2015)
12. Nickel, M., Kiela, D.: Poincaré embeddings for learning hierarchical representations. In: Advances in Neural Information Processing Systems, pp. 6338–6347 (2017). http://papers.nips.cc/paper/7213-poincare-embeddings-for-learning-hierarchical-representations.pdf
13. Nickel, M., Kiela, D.: Learning continuous hierarchies in the Lorentz model of hyperbolic geometry. In: International Conference on Machine Learning, pp. 3776–3785 (2018). https://arxiv.org/abs/1806.03417
14. Ollivier, Y.: Ricci curvature of Markov chains on metric spaces. J. Funct. Anal. **256**(3), 810–864 (2009). https://doi.org/10.1016/j.jfa.2008.11.001
15. O'Neill, B.: Semi-Riemannian Geometry with Applications to Relativity, vol. 103. Academic Press, Cambridge (1983)
16. Perozzi, B., Al-Rfou, R., Skiena, S.: DeepWalk: online learning of social representations. In: Proceedings of the 20th ACM SIGKDD International Conference on Knowledge Discovery and Data Mining, pp. 701–710. ACM (2014)
17. Sala, F., De Sa, C., Gu, A., Re, C.: Representation tradeoffs for hyperbolic embeddings. In: International Conference on Machine Learning, pp. 4457–4466 (2018)
18. Samal, A., Sreejith, R., Gu, J., Liu, S., Saucan, E., Jost, J.: Comparative analysis of two discretizations of Ricci curvature for complex networks. Sci. Rep. **8**(1), 8650 (2018). https://doi.org/10.1038/s41598-018-27001-3

19. Sarkar, R.: Low distortion delaunay embedding of trees in hyperbolic plane. In: van Kreveld, M., Speckmann, B. (eds.) GD 2011. LNCS, vol. 7034, pp. 355–366. Springer, Heidelberg (2012). https://doi.org/10.1007/978-3-642-25878-7_34
20. Sreejith, R., Mohanraj, K., Jost, J., Saucan, E., Samal, A.: Forman curvature for complex networks. J. Stat. Mech.: Theory Exp. **2016**(6), 063206 (2016). https://doi.org/10.1088/1742-5468/2016/06/063206
21. Weber, M., Saucan, E., Jost, J.: Coarse geometry of evolving networks. J. Complex Netw. **6**(5), 706–732 (2017). https://doi.org/10.1093/comnet/cnx049

Information Diffusion in Complex Networks: A Model Based on Hypergraphs and Its Analysis

Alessia Antelmi[1(✉)], Gennaro Cordasco[2], Carmine Spagnuolo[1],
and Przemysław Szufel[3]

[1] Dipartimento di Informatica, Università degli Studi di Salerno, Fisciano, Italy
{aantelmi,cspagnuolo}@unisa.it
[2] Dipartimento di Psicologia, Università degli Studi della Campania
"Luigi Vanvitelli", Caserta, Italy
gennaro.cordasco@unicampania.it
[3] SGH Warsaw School of Economics, Warsaw, Poland
pszufe@sgh.waw.pl

Abstract. This work introduces the problem of social influence diffusion in complex networks, where vertices are linked not only through simple pairwise relationships to other nodes but with groups of nodes of arbitrary size. A challenging problem that arises in this domain is to determine a small subset of nodes S (*a target-set*) able to spread their influence in the whole network. This problem has been formalized and studied in different ways, and many viable solutions have been found for graphs. These have been applied to study several phenomena in research fields such as social, economic, biological, and physical sciences.

In this contribution, we investigated the *social influence problem* on hypergraphs. As hypergraphs are mathematical structures generalization of graphs, they can naturally model the many-to-many relationships characterizing a complex network. Given a network represented by a hypergraph $H = (V, E)$, we consider a dynamic influence diffusion process on H, evolving as follows. At the beginning of the process, the nodes in a given set $S \subseteq V$ are influenced. Then, at each iteration, the influenced hyperedges set is augmented by all hyperedges having a sufficiently large number of influenced nodes. Consequently, the set of influenced nodes is extended by all the nodes contained in a sufficiently large number of already influenced hyperedges. The process terminates when no new nodes can be influenced.

The so defined problem is an inherent chicken-and-egg question as nodes are influenced by groups of other nodes (or hyperedges), while hyperedges (or group of nodes) are influenced by the nodes they contain. In this paper, we provide a formal definition of the influence diffusion problem on hypergraphs. We propose a set of greedy-based heuristic strategies for finding the minimum influence target set, and we present an in-depth analysis of their performance on several classes of random

The research is financed by NAWA—The Polish National Agency for Academic Exchange.

© Springer Nature Switzerland AG 2020
B. Kamiński et al. (Eds.): WAW 2020, LNCS 12091, pp. 36–51, 2020.
https://doi.org/10.1007/978-3-030-48478-1_3

hypergraphs. Furthermore, we describe an experiment on a real use-case, based on the character co-occurrences network of the Game-of-Thrones TV Series.

Keywords: Influence diffusion · Target set selection · Random hypergraphs · Social network

1   Introduction

The current research on social networks is focusing on modeling community structures to examine how and to what extent relationships between humans or entities are the cause of complex emergent phenomena [21,40]. In the past decades, graphs have played an essential role in the modeling and analysis of large-scale online social networks (OSNs) [8,35], such as Facebook, Twitter or Linkedin, as well as for studying biological [27,38] or economic systems [24,41]. Adopting graphs to model these networks assumes the existence of only binary relationships between nodes. However, in many cases, complex networks are characterized by more articulated interactions. For instance, communication networks, reviewing activities, money transactions, geographical tracking, and many other scenarios are governed by many-to-many relationships. For a more clarifying example, we can consider the network built upon email exchanges between some users. In this context, the object *email* can be modeled as a relation involving a group of users. Thus, in this case, nodes of the network represent the persons, while the edges of the network incorporate a sub-set of them – i.e., all email receivers. It is worth noting that if we represent this scenario with a graph, we lose the information about which users are receivers of the same emails. This approach, combined with grouping messages having the same title, can be used for anomaly and spam detection in electronic communication [39]. Recently, hypergraphs have been exploited as a tool for modeling complex networks. Being a generalization of graphs, where a (hyper)edge is a relationship among an arbitrary number of nodes, they can naturally define many-to-many relations between groups of objects, such as domain names and IP addresses [30].

This research constitutes a relatively new area investigated in several recent works [16,33]. A well-known problem in the field of network analysis is the question of social influence maximization, which aims to identify the set of nodes able to spread information in the whole network. However, little research on this topic does take into account many-to-many relationships existing in a complex network. Social influence [13,18] is the process by which each individual change its behavior or adapt its opinions, according to the interactions with other people. With this aim in mind, it is crucial to notice that this process is a fundamental aspect in many fields, such as *viral marketing* [11,22], in which the information diffusion process is used to attract people to adopt products or ideas. According to Lately [32], *"the traditional broadcast model of advertising-one-way, one-to-many, read-only is increasingly being superseded by a vision of marketing that wants and expects, consumers to spread the word themselves"*. The major contributions of this paper are summarized as follows.

- We formally define the dynamic social influence problem on hypergraphs, and we present a variant of the target set problem, first presented in [31], suitable for networks involving many-to-many relationships.
- We introduce four random hypergraphs generative algorithms to build i) random hypergraphs (without any constraint); ii) k-uniform hypergraphs (where each hyperedge has size k); iii) d-regular hypergraphs (where each node has degree d); and iv) hypergraphs with the preferential attachment rule [5].
- We propose three greedy-based heuristics for finding the minimum influence target set on hypergraphs that eventually will influence the whole network.
- We present an evaluation of the proposed algorithms on a set of random hypergraphs, varying the random properties of the networks, and results on a real use-case, based on the network induced by the co-occurrences of characters in the Game-of-Thrones TV Series.

Outline of the Paper. The paper is organized as follows. In Sect. 2, we define the minimum target set problem on hypergraphs, representing networks defined by many-to-many relationships. Furthermore, we describe four generating models of random hypergraphs. Section 3 reviews some relevant literature about the social influence problem and its applications. In Sect. 4, we describe our proposed greedy-based heuristics to solve the social influence problem. Section 5 presents our experiments, and we also discuss results on a real use-case. Finally, Sect. 6 details the conclusion and future work.

2   Background

2.1   Hypergraphs

A hypergraph is an ordered pair $H = (V, E)$ where V is the set of nodes or vertices, which refers to a set of objects, and E is the set of (hyper)edges. Each hyperedge is a non-empty subset of vertices; i.e., $E \subseteq 2^V \setminus \{\emptyset\}$, where 2^V is the power set of V. In this paper, we indicate with $n = |V|$ the number of nodes in V, and with $m = |E|$ the number of hyperedges in E, respectively. A graph is a hypergraph, where each hyperedge is a two element subset of V; in other words, a hypergraph $G = (V, E)$ is a graph if $E \subseteq \binom{V}{2} \subseteq 2^V \setminus \{\emptyset\}$. For a hypergraph H, a two-section representation $[H]_2$ can be obtained by connecting two nodes in the graph $[H]_2$ if and only if they are in the same hyperedge of H [9]. As a result, each hyperedge from H occurs as a complete graph in $[H]_2$. In this work, we considered the weighted $[H]_2$ of H, which assumes that the weight of an edge corresponds to the number of hyperedges containing both the edge endpoints.

2.2   Dynamic Social Influence Diffusion on Hypergraphs

Given a network represented by a hypergraph $H = (V, E)$, we consider a dynamic influence diffusion process on H, which evolves in discrete steps as follows. In the beginning, the nodes in a given set $S \subseteq V$ are influenced. Then, at each iteration:

1. the influenced hyperedges set is augmented by all edges which have a sufficiently large number of influenced nodes;
2. consequently, the set of influenced nodes is augmented by all the nodes which
   have a sufficiently large number of already influenced edges.

The process ends if no new nodes can be influenced.

Formally, let $H = (V, E)$ be a hypergraph. For each $v \in V$, we denote with
$E(v) \subseteq E$ the set of edges that contains v and with $d(v) = |E(v)|$ the degree of
v. Analogously, for each $e \in E$, we denote with $V(e) \subseteq V$ the set of nodes in e
and with $k(e) = |V(e)|$ the cardinality of e. Let $t_V : V \to \mathbb{N} = \{0, 1, \dots\}$ and
$t_E : E \to \mathbb{N} = \{0, 1, \dots\}$ be two functions assigning thresholds to the vertices
and to the hyperedges, respectively. For each node $v \in V$ (resp. $e \in E$), the
value $t_V(v)$ (resp. $t_E(e)$) quantifies how hard it is to influence node v (edge e),
in the sense that easy-to-influence elements of the network have "low" threshold
values, and hard-to-influence elements have "high" threshold values.

Definition 1. *Let $H = (V, E)$ be a hypergraph with threshold functions $t_V :$
$V \longrightarrow \mathbb{N}$ and $t_E : E \longrightarrow \mathbb{N}$, and $S \subseteq V$. An information diffusion process in H,
starting with a seed $S \subseteq V$, is a sequence*

$$I_V[S, 0] \subseteq I_V[S, 1] \subseteq \dots \subseteq I_V[S, \ell] \subseteq \dots \subseteq V$$

of vertex subsets, with $I_V[S, 0] = S$, and

$$I_E[S, 0] \subseteq I_E[S, 1] \subseteq \dots \subseteq I_E[S, \ell] \subseteq \dots \subseteq E$$

of edge subsets, with $I_E[S, 0] = \emptyset$ and and such that for all $\ell > 0$

$$I_E[S, \ell] = I_E[S, \ell - 1] \bigcup \left\{ e \in E : |V(e) \cap I_V[S, \ell - 1]| \geq t_E(e) \right\}$$

$$I_V[S, \ell] = I_V[S, \ell - 1] \bigcup \left\{ v \in V : |E(v) \cap I_E[S, \ell]| \geq t_V(v) \right\}$$

*A **target set** for H is a seed set $S \subseteq V$ that will eventually influence the whole
network (i.e., $I_V[S, r] = V$ for some $r \geq 0$).*

We indicate the above information diffusion process on H with

$$I_V[S], I_E[S] = \Phi(H, S, t_V, t_E),$$

where, $I_V[S] \subseteq V$ is the set of influenced vertices ($I_V[S] = I_V[S, r]$), and
$I_E[S] \subseteq E$ is the set of influenced hyperedges. t_V and t_E denote the thresholds
functions for nodes and hyperedges, respectively.

Example 1. Consider the hypergraph H in Fig. 1. The nodes are depicted as
an oval shape. The number on the top represents the node identifier; on the
bottom, its threshold value is shown. The hyperedge threshold value is drawn as
a black half oval shape. The hyperedge identifier is depicted inside the hyperedge.
Finally, influenced nodes are drawn in gray. Influenced hyperedges are shaped

using a gray dotted line. Given a possible seed set S for H equal to $\{v_1, v_4\}$, the information diffusion process evolves as follows.

$$I_E[S, 0] = \emptyset, \quad I_V[S, 0] = S = \{v_1, v_4\}$$
$$I_E[S, 1] = \{e_2\}, \quad I_V[S, 1] = \{v_1, v_3, v_4\}$$
$$I_E[S, 2] = \{e_2, e_3\}, \quad I_V[S, 2] = \{v_1, v_3, v_4, v_5\}$$
$$I_E[S, 3] = \{e_1, e_2, e_3\}, \quad I_V[S, 3] = \{v_1, v_2, v_3, v_4, v_5\} = V.$$

Hence, S is a target set for H.

The problem examined in this paper is defined as follows:

Problem 1. Diffusion on Hypergraphs—DoH
Instance: $H = (V, E)$, thresholds $t_V : V \to N_0$ and $t_E : E \to N_0$.
Problem: Find a seed set $S \subseteq V$ of minimum size such that $I_V[S] = V$.

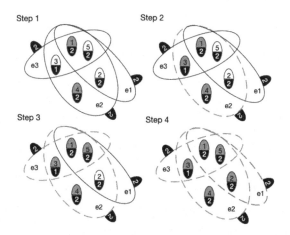

Fig. 1. An example of social influence diffusion process on $H = (V, E)$.

2.3 Models for Random Hypergraphs

In this work, we investigate information diffusion processes on complex networks by exploiting random hypergraphs. Here, we describe four generative models, characterized according to the structural proprieties of the computed random hypergraph (for example, hypergraphs with a fixed degree of nodes).

1. *Random model.* It generates a hypergraph without any structural property constraint. Given two integer parameters n and m (the number of nodes and hyperedges, respectively), the algorithm computes - for each hyperedge $he = 1, \ldots, m$ - a random number $s \in [1, n]$ (i.e. the hyperedge size). Then, the algorithm selects uniformly at random s vertices from V to add in he.
2. *K-uniform model.* It generates a k-uniform hypergraph, which is a hypergraph where each hyperedge has a size of k. The algorithm proceeds as the *random* model, but forcing the size of each hyperedge equal to k.

3. *D-regular model.* It generates a d-regular hypergraph, where each node has degree d. The algorithm exploits the k-uniform approach, described above, to build a d-regular hypergraph H having m nodes and n edges. It then returns the hypergraph H^*, dual of H.
4. *Preferential-attachment model.* It generates a hypergraph with a preferential attachment rule between nodes, as described in [5]. The algorithm starts with a fully-random graph with 5 nodes and 5 edges. It then iteratively adds a node or an edge, according to a given parameter p, defining the probability of creating a new node or a new hyperedge. In detail, the connections with the new node or hyperedge are generated according to a preferential attachment policy [5]. We slightly changed the algorithm to avoid repetitions in the hyperedges.

3   Related Work

The Social Influence Diffusion Problem. Previous research showed the importance of the target set selection (TSS) problem to study the social influence diffusion in networks. The TSS problem aims to select k initially-influenced seed users to maximize the expected number of eventually-influenced users. In other words, the objective is to find a subset of nodes in the network that, once active, can activate all the nodes of the network under the linear threshold (LT) influence propagation model. According to the LT model, a user v becomes active when the sum of influences of its neighbors in the networks reaches a specific threshold $t(v)$ [17]. Given its importance in the context of influence spread in both the online (social networks) and offline (word-of-mouth) worlds, the TSS problem is extensively studied on graphs. Kempe et al. [31] first analyzed the problem in networks with randomly chosen thresholds. Chen [12] studied the minimization problem of finding the smallest target set able to influence the whole network built with fixed arbitrary thresholds. Furthermore, Chen proved a strong inapproximability result that makes unlikely the existence of an algorithm for the TSS problem on graphs (2-uniform hypergraphs) with an approximation factor better than $O(2^{\log^{1-\epsilon} |V|})$. Cordasco et al. [16] presented an algorithm for the TSS always producing an optimal solution (i.e., a minimum size subset of nodes that influence the whole network) in case the network is either a tree, a cycle, or a complete graph.

Considering that researchers started focusing on hypergraphs only in the last decade, little or no literature exists on the TSS problem on hypergraphs. Zhu et al. [42] deal with the problem of social influence maximization in social networks. They model the crowd influence as a hyperedge $e = (H_e, v)$ with weight $0 \le P_e \le 1$, where H_e is the head node-set and v is the tail node, meaning that v will be activated by H_e with probability P_e only after each node in H_e is activated. Their proposed algorithm selects k initially-influenced seed users in a directed hypergraph $G = (V, E, P)$, maximizing the expected number of eventually-influenced users. Another stochastic diffusion process in which information diffusion can occur through interactions in groups of different sizes is described by Iacopini et al. [28].

Our study addresses the social influence diffusion problem on networks characterized by many-to-many relationships, using undirected hypergraphs, which allow modeling more kinds of real-world use cases, such as social networks like Facebook or Yelp. Furthermore, in our work, we adopted a linear thresholds model, investigating different threshold values for nodes and hyperedges. We also present a deterministic model which is more suitable for real use-cases.

Random Hypergraphs Generation. The foundations of random graph theory lie in a seminal paper by Erdős and Rényi [20]. However, several models have been developed that make it possible to generate random graphs having desired topological properties to better mimic the real world. The Barabási-Albert models *rich-get-richer* phenomena. On the other hand Watts-Strogat *small-world* model is useful for representation of social networks. Random graph structures have proved to be a useful concept in many disciplines. Still, more complex mathematical tools are needed to comprehensively and accurately model many real-world complex networks [5]. The study of random hypergraph models has its origin from work by Erdős and Bollobas [7], which presents an analogous to the Erdős-Rényi random graph model. In the following years, researchers focused on analyzing several properties of this model [1,15,19,23]. Wang et al. [29] first defined a preferential attachment model for hypergraphs, with vertex arrival events and constant-size hyperedges. Starting from this model and its limitations, Avin et al. [5] proposed a preferential attachment model generating hypergraphs with hyperedges of arbitrary size, allowing cycles and non-uniformity. In particular, they extended the Chung-Lu preferential attachment model proposed for graphs [14].

4   Finding the Minimum Target Set on Hypergraphs

In this Section, we discuss three greedy-based heuristics for the *DoH* problem (see Sect. 2.2), i.e., finding the minimum influence target set $S \subseteq V$ of a hypergraph $H = (V, E)$ able to influence the whole network. A simple greedy strategy may be selecting - at each iteration - the nodes in descending order by their degree until the current set can influence the whole network. We refer to this approach with the label *StaticGreedy*. It enables us to compute the set S by exploiting a binary search strategy detailed in Algorithm 1. As described in Sect. 2.2, we indicate the diffusion process on H with $\Phi(H, S)$, and we denote with $I_V[S] \subseteq V$ and $I_E[S] \subseteq E$ the end set of influenced nodes and hyperedges, respectively.

A dynamic approach, referred to as *DynamicGreedy*, is listed in Algorithm 2. In this heuristic, all nodes are added to the candidates set U. At each stage, the node of the maximum degree is added to S and removed from U. At this point, some nodes and/or hyperedges become infected. The algorithm simulates the diffusion process, and influenced edges are pruned from the network. The degree of nodes $(\delta(v))$ is updated accordingly.

Algorithm 1. $StaticGreedy(H = (V, E), t_V, t_E)$

1: Let $\sigma(V)$ be the list of nodes in descending order of their degree $d(v)$.
2: $left = 1$, $right = |V|$
3: **while** $left < right$ **do**   ▷ Binary Search
4:     $mid = \left\lceil \frac{left+right}{2} \right\rceil$
5:     $I_V[S], I_E[S] = \Phi(H, \sigma_{mid}, t_V, t_E)$ ▷ σ_i denotes the set containing the first i nodes in the order $\sigma(V)$;
6:     **if** $I_V[S]! = V$ **then**
7:         $left = mid$
8:     **else**
9:         $right = mid - 1$
10: **return** $S = \sigma_{left+1}$

Algorithm 2. $DynamicGreedy(H = (V, E), t_V, t_E)$

1: $S = \emptyset$, $U = V$, $E' = E$
2: **for** $u \in U$ **do**
3:     $\delta(u) = d(u)$
4: **while** $U \neq \emptyset$ **do**
5:     $v = argmax_{u \in U}\, \delta(v)$
6:     $U = U \setminus \{v\}$
7:     $S = S \cup \{v\}$
8:     $I_V[S], I_E[S] = \Phi(H, S, t_V, t_E)$
9:     **if** $I_V[S] = V$ **then**
10:         break;
11:     $E' = E - I_E[S]$
12:     **for** $u \in U$ **do**
13:         $\delta(u) = |E(u) \cap E'|$ ▷ $\delta(u)$ denotes the degree of u in $H = (V, E')$.
14: **return** S

Given the *DynamicGreedy* algorithm, we have designed a similar heuristic, named *DynamicGreedy$_{[H]_2}$*, and listed in Algorithm 2. In this heuristic, we compute the degree of the nodes on the $[H]_2$ of the residual hypergraph H^i of H. H^i is the hypergraph obtained removing all hyperedges already influenced by the nodes in S at stage i.

Algorithm 3. $DynamicGreedy_{[H]_2}(H(V, E), t_V, t_E)$

1: $S = \emptyset$, $U = V$, $E' = E$, $[H]_2 = 2Section(H(V, E))$
2: **while** $U \neq \emptyset$ **do**
3:     $v = argmax_{u \in U}\, d_{[H]_2}(v)$ ▷ $d_{[H]_2}(v)$ denotes the degree of v in $[H]_2$.
4:     $U = U \setminus \{v\}$
5:     $S = S \cup \{v\}$
6:     $I_V[S], I_E[S] = \Phi(H, S, t_V, t_E)$
7:     **if** $I_V[S] = V$ **then**
8:         break;
9:     $E' = E - I_E[S]$
10:     $[H]_2 = 2Section(H(V, E'))$
11: **return** S

---

5   Experiments

We present experiments on the three greedy-based heuristics discussed in Sect. 4. We investigated two classes of experiments; we evaluated the proposed heuristics on random networks, and on a real use-case by exploiting the co-occurrences network of the TV Series Game-of-Thrones.

5.1   Random Networks

We performed three experimental scenarios for the case of random hypergraphs. In the first and second scenarios, we fixed the node threshold to a random value between 1 and its degree. In the last scenario, each node threshold varies proportionally - from 0.1 to 0.9 - to the degree of the node. In particular, in the first scenario, we run the heuristics on random hypergraphs with no structural proprieties generated with the *random* model and hypergraphs generated with the *preferential-attachment* rule. We ranged the hypergraph size, using $[100, 200, 400, 800]$ nodes and hyperedges. In the second scenario, we experimented the heuristics on *k-uniform* and *d-regular* random hypergraphs, ranging the value of k and d in $[10, 20, 40, 80]$. In the third and last scenario, we generated a random hypergraph of fixed size ($n = m = 500$) with all generative models. We fixed both $k = 80$ and $d = 80$, for the *k-uniform* and *d-regular* random hypergraphs, respectively. In all experiments, we set each hyperedge activation threshold proportional to its degree scaled of factor 0.5 (majority policy). We executed each experiment 48 times. We implement all heuristics and experiments in *Julia* language, by exploiting the library `SimpleHypergraphs.jl` [3]. The Julia code used in the paper is available at the following public GitHub repository[1].

*Scenario 1—**Increasing H Size, Random Thresholds.*** Figure 2 shows the results obtained on random hypergraphs - with different sizes - generated by the *random* and *preferential-attachment* models. On the y-axis, we report the size of the influence target set S; on the x-axis, the hypergraph size ($n = m$). The

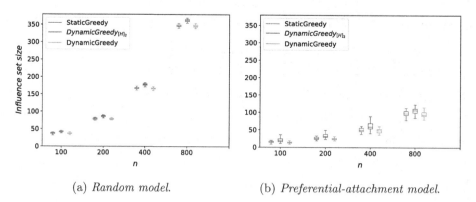

(a) *Random model.*   (b) *Preferential-attachment model.*

Fig. 2. Experiments on random hypergraphs $H = (V, E)$, generated with the random and preferential-attachment models, varying the degree of nodes and hyperedges ($n = m$). For each node, the threshold is fixed to a random value between 1 and the node degree. A fixed threshold to 0.5 is used for hyperedges.

---

[1] https://github.com/pszufe/LTMSim.jl.

DynamicGreedy heuristic achieves the best average performance. However, as shown in Fig. 2a, there is not a significant difference between the three strategies. On the other hand, the *DynamicGreedy* heuristic significantly outperforms the others in the case of the *preferential-attachment* scenario, as shown in Fig. 2b.

*Scenario 2—**Uniform and Regular H, Random Thresholds**.* Figure 3 shows the results obtained on random hypergraphs generated by the *k-uniform* and *d-regular* models. On the *x*-axis, we show values for *k* and *d*. As shown in Fig. 3a, the *DynamicGreedy* strategy achieves better results for random *k-uniform* hypergraphs, especially in the case of large values of *k*. Figure 3b depicts the results for random *d-regular* hypergraphs. By increasing the size of *d*, there is no significant difference between the heuristics, even if for small values of *d*, their results exhibit a more significant variance. It is worth discussing the interesting - even though not so surprising - outcomes revealed by the comparison of the results obtained in the *k-uniform* and *d-regular* experiments. In general, the *k-uniform* hypergraphs require a target set of smaller size compared to *d-regular* hypergraphs.

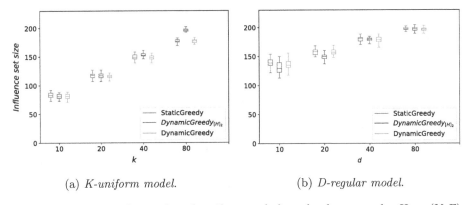

(a) *K-uniform model.*   (b) *D-regular model.*

Fig. 3. Experiments for random *k*-uniform and *d*-regular hypergraphs $H = (V, E)$, with a fixed hypergraph size $n = m = 500$. For each node, the threshold is fixed to a random value between 1 and the node degree. A fixed threshold to 0.5 is used for hyperedges.

*Scenario 3—**Varying Node Thresholds Proportionally to Their Degree**.* Fig. 4 outlines the results obtained on a hypergraph *H* of fixed size $n = m = 500$, generated by each random model. We ranged nodes activation thresholds proportionally to their degree size from 0.1 to 0.9, and we fixed the hyperedges activation threshold proportionally to 0.5. The heuristics achieve almost the same performance in the case of a completely *random* graph (Fig. 4a) and a *d-regular* (Fig. 4d) hypergraph. Results obtained from the *preferential-attachment* (Fig. 4b) and *k-uniform* (Fig. 4c) models are more attractive. In both experiments, *DynamicGreedy*$_{[H]_2}$ exhibits the worst results compared to the other two heuristics. Interestingly, the preferential-attachment case

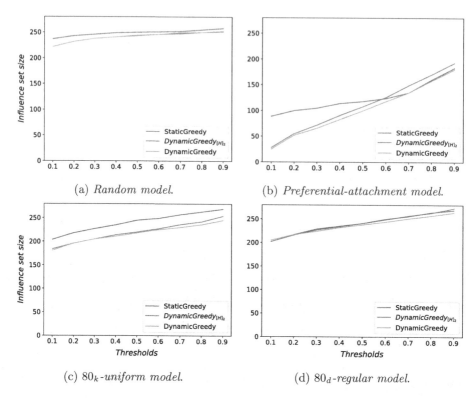

(a) *Random model.* (b) *Preferential-attachment model.*

(c) 80_k-*uniform model.* (d) 80_d-*regular model.*

Fig. 4. Experiments for random hypergraphs $H = (V, E)$, of size $n = m = 500$, considering all random generative models. For each node, the threshold varies proportionally - from 0.1 to 0.9 - to the degree of the node. A fixed threshold to 0.5 is used for hyperedges. The value of k and d for the k-uniform and d-regular hypergraphs is set to 80.

exhibits unusual behavior. When the thresholds are small, the performance of $DynamicGreedy_{[H]_2}$ is poor, but for larger values, its performance improves and is very close to the $DynamicGreedy$ heuristic. As a result of using high threshold values, it is hard to trigger an information cascade in the network as, in this case, the influence diffusion process behaves more like a domination process. In general, this makes the problem easier to face.

5.2   Game-of-Thrones TV Series Network

Game of Thrones [25] (GoT) is the screen adaption of the series of fantasy novels *A Song of Ice and Fire*, written by George R.R. Martin. Created by D. Benioff and D.B. Weiss for the American television network HBO, the American fantasy drama TV series has attracted a record viewership and has a broad, active, and international fan base—according to Wikipedia[2]. This enthusiasm has led

---

[2] https://en.wikipedia.org/wiki/Game_of_Thrones.

the intricate world of GoT to be a profoundly immersive entertainment experience [4]. Both the academic community and industries took the opportunity to study not only complex dynamics within the GoT storyline [6], but also how viewers engage with the GoT world on social media [2,26,37], or how the novel itself is a portrait of real-world dynamics [10,34,36].

In this experiment, we exploited GoT season episodes data from the dataset *Game of Thrones Datasets and Visualizations*, available at the following GitHub repository[3]. Specifically, we used information describing each episode scenes. They contain - for each scene - start, end, location, and a list of characters performing in it. Table 1 reports some necessary information about the number of episodes, scenes, and characters per GoT season. A more detailed description of the dataset is available on the dataset GitHub repository.

The GoT Network—H_{GoT}. We modeled the GoT network using a hypergraph H_{got}, considering the characters co-occurrences within scenes per each season. The vertices of H_{got} represents the 577 GoT characters. Each hyperedge of H_{got}, therefore, link together all characters that have acted in the same scene together. The total number of considered scenes was 4165. Figure 5 presents the hyperedges size distribution of H_{got}. It shows a typical power-law distribution, where few scenes assemble a considerable number of characters. In contrast, many others focus on few or no characters.

Table 1. Some GoT dataset numbers.

Season	Episodes	Scenes	Characters
1	10	286	125
2	10	468	137
3	10	470	137
4	10	517	152
5	10	508	175
6	10	577	208
7	7	468	75
8	6	871	66

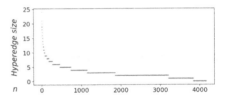

Fig. 5. H_{GoT} hyperedges distribution.

Influencing the GoT Network. We performed two experiments on the GoT network, aiming at evaluating the performance of the heuristics in minimizing the number of nodes (or characters) to influence. In Fig. 6, we detail the performance of each heuristic both in the case of random threshold values for each node (Fig. 6a), and in the case of proportional threshold values (Fig. 6a). The *DynamicGreedy* and *DynamicGreedy*$_{[H]_2}$ provide similar results requiring a

---

[3] Game of Thrones Datasets and Visualizations. https://github.com/jeffreylancaster/game-of-thrones by Jeffrey Lancaster.

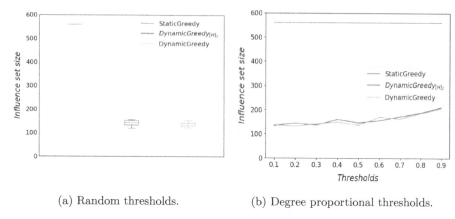

(a) Random thresholds.   (b) Degree proportional thresholds.

Fig. 6. Experiments for the GoT network using a) random and b) proportional nodes thresholds values. A fixed threshold to 0.5 is used for hyperedges.

seed set of about 120 nodes on average. The second case shows the same trend, and they can find reasonable solutions and achieve, in the worst-case (0.9), a target set of size about 30% of V. On the contrary, *StaticGreedy* provides a target set almost equal to V for each threshold.

6   Conclusions and Future Work

This paper faces the social influence diffusion process in complex networks, exploiting the hypergraph structure. We propose a formulation of the dynamic influence diffusion on hypergraphs, referred to as the *Diffusion on Hypergraphs* (DoH) problem. The so-defined problem on hypergraphs differs from the correspondent on graphs, as it introduces the influence propagation also on the network connections, i.e., the hyperedges (which denote groups of related nodes).

A challenging problem arising in this domain is to determine a small subset of nodes S (*a target-set*) able to spread their influence in the whole network. We present three greedy-based heuristics to solve this problem on hypergraphs, considering either the degree of nodes in the hypergraph H or in the two-section view $[H]_2$ of H, and selecting the nodes according to static or dynamic policies. We provided an exhaustive investigation of their performance on a bunch of random networks and a real use-case based on the character co-occurrences in the GoT TV series. We observed that the *DynamicGreedy* heuristic achieved the best results in the case of random networks. In the real use-case of the GoT network, experiments highlighted that dynamically selecting the nodes (according to their degree in the residual hypergraph) to add to the target set results in a more efficient solution compared to a static approach. Furthermore, for the GoT network, we also noticed that the dynamic greedy-based heuristics (*DynamicGreedy* and *DynamicGreedy*$_{[H]_2}$) provided a good seed set when choosing an initial set of size at most 30% of V.

As future work, we plan to investigate more efficient algorithms and approaches for the *DoH* problem. Furthermore, we aim to experiments with the proposed strategies on real-world datasets, such as a *Twitter* social network built upon tweet hashtags or user reviews from the *Yelp.com* dataset. Results are encouraging, and further investigation is needed to explore the social influence diffusion problem on hypergraphs as it might shed light on complex social phenomena, like fake news sharing in online social networks.

References

1. Aksoy, S.G., Kolda, T.G., Pinar, A.: Measuring and modeling bipartite graphs with community structure. J. Complex Netw. **5**(4), 581–603 (2017)
2. Antelmi, A., Breslin, J., Young, K.: Understanding user engagement with entertainment media: a case study of the twitter behaviour of Game of Thrones (GoT) fans. In: 2018 IEEE Games, Entertainment, Media Conference (GEM) (2018)
3. Antelmi, A., et al.: SimpleHypergraphs.jl–novel software framework for modelling and analysis of hypergraphs. In: Algorithms and Models for the Web Graph, pp. 115–129 (2019)
4. Askwith, I.D.: Television 2.0: reconceptualizing TV as an engagement medium (2007). https://dspace.mit.edu/handle/1721.1/41243
5. Avin, C., Lotker, Z., Peleg, D.: Random preferential attachment hypergraphs. Computer Science 23 (2015)
6. Beveridge, A., Shan, J.: Network of thrones. Math. Horiz. **23**, 4 (2016)
7. Bollobas, B., Erdös, P.: Cliques in random graphs. Math. Proc. Cambridge Philos. Soc. **80**(3), 419–427 (1976)
8. Bonato, A., Eikmeier, N., Gleich, D.F., Malik, R.: Dynamic competition networks: detecting alliances and leaders. In: Algorithms and Models for the Web Graph (2018)
9. Bretto, A.: Hypergraph Theory: An Introduction. Springer, Heidelberg (2013). https://doi.org/10.1007/978-3-319-00080-0
10. Bruno, L., Miriam, H.: The house of black and white: identities of color and power relations in the game of thrones. Revista Nós **4**(2), 161–182 (2019)
11. Cautis, B., Maniu, S., Tziortziotis, N.: Adaptive influence maximization. In: Proceedings of the 25th ACM SIGKDD International Conference on Knowledge Discovery & Data Mining, pp. 3185–3186 (2019)
12. Chen, N.: On the approximability of influence in social networks. SIAM J. Discrete Math. **23**(3), 1400–1415 (2009)
13. Chen, W., Lakshmanan, L.V.S., Castillo, C.: Information and Influence Propagation in Social Networks. Morgan & Claypool Publishers, San Rafael (2013)
14. Chung, F.R.K., Lu, L.: Complex graphs and networks. In: CBMS Regional Conference Series in Mathematics (2006)
15. Cooper, C., Frieze, A., Molloy, M., Reed, B.: Perfect matchings in random r-regular, s-uniform hypergraphs. Comb. Probab. Comput. **5**(1), 1–14 (1996)
16. Cordasco, G., Gargano, L., Mecchia, M., Rescigno, A.A., Vaccaro, U.: Discovering small target sets in social networks: a fast and effective algorithm. Algorithmica **80**(6), 1804–1833 (2018)
17. Cordasco, G., Gargano, L., Rescigno, A.A.: On finding small sets that influence large networks. Soc. Netw. Anal. Min. **6**(1), 1–20 (2016). https://doi.org/10.1007/s13278-016-0408-z

18. Easley, D., Kleinberg, J.: Networks, Crowds, and Markets: Reasoning About a Highly Connected World. Cambridge University Press, Cambridge (2010)
19. Ellis, D., Linial, N.: On regular hypergraphs of high girth. Electron. J. Comb. **21**, 1 (2014)
20. Erdős, P., Rényi, A.: On the evolution of random graphs. In: Publication of the Mathematical Institute of Hungarian Academy Of Science, pp. 17–61 (1960)
21. Fani, H., Jiang, E., Bagheri, E., Al-Obeidat, F., Du, W., Kargar, M.: User community detection via embedding of social network structure and temporal content. Inf. Process. Manage. **57**, 2 (2020)
22. Ghayoori, A., Nagi, R.: Seed investment bounds for viral marketing under generalized diffusion. In: Proceedings of the 2019 IEEE/ACM International Conference on Advances in Social Networks Analysis and Mining, pp. 95–100 (2019)
23. Ghoshal, G., Zlatic, V., Caldarelli, G., Newman, M.E.J.: Random hypergraphs and their applications. Phys. Rev. E: Stat. Nonlin. Soft Matter Phys. **79**, 066118 (2009)
24. Göbel, M., Araújo, T.: A network structure analysis of economic crises. In: Complex Networks and Their Applications VIII, pp. 547–560 (2020)
25. HBO: Game of Thrones (2019). https://www.hbo.com/game-of-thrones
26. Héctor, J.P., Rainer, R.: On Jon Snow's death: plot twist and global fandom in Game of Thrones. Cult. Psychol. (2019). https://doi.org/10.1177/1354067X19845062
27. Hossain, M., Khan, A., Uddin, S.: Understanding the progression of congestive heart failure of type 2 diabetes patient using disease network and hospital claim data. In: Complex Networks and Their Applications, VIII (2020)
28. Iacopini, I., Petri, G., Barrat, A., Latora, V.: Simplicial models of social contagion. Nat. Commun. **10**, 2485 (2019)
29. Jian-Wei, W., Li-Li, R., Qiu-Hong, R., Ji-Yong, Z.: Evolving hypernetwork model. Phys. Condens. Matter **77**, 493–498 (2010)
30. Joslyn, C., et al.: High performance hypergraph analytics of domain name system relationships. In: HICSS Symposium on Cybersecurity Big Data Analytics (2019)
31. Kempe, D., Kleinberg, J., Tardos, E.: Maximizing the spread of influence through a social network. In: Proceedings of the Ninth ACM SIGKDD International Conference on Knowledge Discovery and Data Mining, pp. 137–146 (2003)
32. Lately, D.: An Army of Eyeballs: The Rise of the Advertisee (2014). https://thebaffler.com/latest/an-army-of-eyeballs. (Online; 2019)
33. Liqing, Q., Jinfeng, Y., Xin, F., Wei, J., Wenwen, G.: Analysis of influence maximization in temporal social networks. IEEE Access **7**, 42052–42062 (2019)
34. Milkoreit, M.: Pop-cultural mobilization: deploying game of thrones to shift us climate change politics. Int. J. Polit. Cult. Soc. **32**(1), 61–82 (2019)
35. Moutidis, I., Williams, H.T.P.: Utilizing complex networks for event detection in heterogeneous high-volume news streams. In: Complex Networks and Their Applications VIII, pp. 659–672 (2020)
36. Muno, W.: "Winter is coming?" *Game of Thrones* and realist thinking. In: Hamenstädt, U. (ed.) The Interplay Between Political Theory and Movies, pp. 135–149. Springer, Cham (2019). https://doi.org/10.1007/978-3-319-90731-4_9
37. Rhodes, R.E., Zehr, E.P.: Fight, flight or finished: forced fitness behaviours in game of thrones. Br. J. Sports Med. **53**(9), 576–580 (2019)
38. Romero, M., Finke, J., Quimbaya, M., Rocha, C.: In-silico gene annotation prediction using the co-expression network structure. In: Complex Networks and Their Applications, VIII (2020)
39. Silva, J., Willett, R.: Hypergraph-based anomaly detection of high-dimensional co-occurrences. IEEE Trans. Pattern Anal. Mach. Intell. **31**(3), 563–569 (2008)

40. Tien, J., Eisenberg, M., Cherng, S., Porter, M.: Online reactions to the 2017 'unite the right' rally in charlottesville: measuring polarization in twitter networks using media followership. Appl. Netw. Sci. **5**, 1 (2020)
41. Verba, M.A.: "Learning hubs" on the global innovation network. In: Complex Networks and Their Applications VIII, pp. 620–632 (2020)
42. Zhu, J., Zhu, J., Ghosh, J., Wu, W., Yuan, J.: Social influence maximization in hypergraph in social networks. IEEE Trans. Netw. Sci. Eng. **6**(4), 801–811 (2019)

A Scalable Unsupervised Framework for Comparing Graph Embeddings

Bogumił Kamiński[1], Paweł Prałat[2], and François Théberge[3(✉)]

[1] Decision Analysis and Support Unit, SGH Warsaw School of Economics,
Warsaw, Poland
bogumil.kaminski@sgh.waw.pl
[2] Department of Mathematics, Ryerson University, Toronto, ON, Canada
pralat@ryerson.ca
[3] Tutte Institute for Mathematics and Computing, Ottawa, ON, Canada
theberge@ieee.org

Abstract. Graph embedding is a transformation of vertices of a graph into a set of vectors. A good embedding should capture the graph topology, vertex-to-vertex relationship, and other relevant information about the graph, its subgraphs, and vertices. If these objectives are achieved, an embedding is a meaningful, understandable, and often compressed representations of a network. Unfortunately, selecting the best embedding is a challenging task and very often requires domain experts.

In the recent paper [1], we propose a "divergence score" that can be assigned to embeddings to help distinguish good ones from bad ones. This general framework provides a tool for an unsupervised graph embedding comparison. The complexity of the original algorithm was quadratic in the number of vertices. It was enough to show that the proposed method is feasible and has practical potential (proof-of-concept). In this paper, we improve the complexity of the original framework and design a scalable approximation algorithm. Moreover, we perform some detailed quality and speed benchmarks.

Keywords: Graph embedding · Geometric Chung-Lu Model

1 Introduction

The study of networks has emerged in diverse disciplines as a means of analyzing complex relational data. Indeed, capturing aspects of a complex system as a graph can bring physical insights and predictive power [2]. Network Geometry is a rapidly developing approach in Network Science [3] which further abstracts the system by modelling the vertices of the network as points in a geometric space. There are many successful examples of this approach that include latent space models, and connections between geometry and network clustering and community structure. Very often, these geometric embeddings naturally correspond to physical space, such as when modelling wireless networks or when networks are

© Crown 2020
B. Kamiński et al. (Eds.): WAW 2020, LNCS 12091, pp. 52–67, 2020.
https://doi.org/10.1007/978-3-030-48478-1_4

embedded in some geographic space. See [4] for more details about applying spatial graphs to model complex networks.

Another important application of geometric graphs is in graph embedding. The idea here is that, for a given network, one tries to embed it in a geometric space by assigning coordinates to each vertex such that nearby vertices are more likely to share an edge than those far from each other. In a good embedding most of the network's edges can be predicted from the coordinates of the vertices. Unfortunately, in the absence of a general-purpose representation for graphs, very often graph embedding requires domain experts to craft features or to use specialized feature selection algorithms. Having said that, there are some graph embedding algorithms that work without any prior or additional information other than graph structure, but these are randomized algorithms that are usually not very stable; that is, the outcome of two applications of the algorithm is often drastically different despite the fact that all the algorithm parameters remain the same.

Consider a graph $G = (V, E)$ on n vertices, and several embeddings of its vertices in some multidimensional spaces (possibly in different dimensions). The main question we try to answer in this paper is: how do we evaluate these embeddings? Which one is the best and should be used? In order to answer these questions, we propose a general framework that assigns the divergence score to each embedding which, in an unsupervised learning fashion, distinguishes good from bad embeddings. In order to benchmark embeddings, we generalize the well-known Chung-Lu random graph model to incorporate geometry. The model is interesting on its own and should be useful for many other problems and tools. In order to test our algorithm, in [1] we experimented with synthetic networks as well as real-world networks, and various embedding algorithms. In this paper, we concentrate on the complexity challenges of the original algorithm and propose a fast approximated algorithm that works very well in practice.

The paper is structured as follows. In Sect. 2, we describe our algorithm for comparing graph embeddings, and we illustrate our approach on one simple graph. The Chung-Lu model is generalized in Sect. 3. In the recent paper [1], we experimented with many datasets and embedding algorithms to show that the framework works well. In this paper, for illustration purposes, we use some of these datasets and their corresponding embeddings. However, due to the space limitation, we do not explain how they are constructed. Interested reader is directed to [1] for more details. Here, we focus on improvements that were required to make an algorithm scalable. The results presented in Sect. 4 for a novel extension of the original algorithm are the main contribution of this paper. We conclude with a discussion on some future directions in Sect. 5.

2   General Framework

Suppose that we are given a graph $G = (V, E)$ on n vertices with the degree distribution $\mathbf{w} = (w_1, \ldots, w_n)$ and an embedding of its vertices to k-dimensional space, $\mathcal{E} : V \to \mathbb{R}^k$. Our goal is to assign a "divergence score" to this embedding.

The lower the score, the better the embedding is. This will allow us to compare several embeddings, possibly in different dimensions.

2.1 Intuition Behind the Algorithm

What do we expect from a good embedding? As already mentioned, in a good embedding, one should be able to predict most of the network's edges from the coordinates of the vertices. Formally, it is natural to expect that if two vertices, say u and v, are far away from each other (that is, $\mathrm{dist}(\mathcal{E}(u), \mathcal{E}(v))$ is relatively large), then the chance they are adjacent in the graph is smaller compared to another pair of vertices that are close to each other. But, of course, in any real-world network there are some sporadic long edges and some vertices that are close to each other are not adjacent. In other words, we do not want to pay attention to local properties such as existence of particular edges (microscopic point of view) but rather evaluate some global properties such as density of some relatively large subsets of vertices (macroscopic point of view). So, how can we evaluate if the global structure is consistent with our expectations and intuition without considering individual pairs?

The approach we take is as follows. We identify dense parts of the graph by running some good graph clustering algorithm. As was illustrated in [1], the choice of graph clustering algorithm is flexible so long as the vertex set is partitioned into clusters such that there are substantially more edges captured within clusters than between them. The clusters that are found will provide the desired macroscopic point of view of the graph. Note that for this task we only use information about the graph G; in particular, we do not use the embedding $\mathcal{E}$ at all. We then consider the graph G from a different point of view. Using the Geometric Chung-Lu (GCL) model that we introduce in this paper especially for this purpose, based on the degree distribution $\mathbf{w}$ and the embedding $\mathcal{E}$, we compute the expected number of edges within each cluster found earlier, as well as between them. The embedding is scored by computing a divergence score between these expected number of edges, and the actual number of edges present in G. Our approach falls into a general and commonly used method of *statistical inference*, in our case applied to the Geometric Chung-Lu model. With these methods, one fits a generative model of a network to observed network data, and the parameters of the fit tell us about the structure of the network in much the same way that fitting a straight line through a set of data points tells us about their slope.

Finally, let us make a comment that not all embeddings proposed in the literature try to capture edges. Some algorithms indeed try to preserve edges whereas others care about some other structural properties; for example, they might try to map together nodes with similar functions. Because of the applications we personally need to deal with require preserving (global) edge densities, our framework favours embeddings that do a good job from that perspective.

2.2   Algorithm

Given a graph $G = (V, E)$, its degree distribution $\mathbf{w}$ on V, and an embedding $\mathcal{E} : V \to \mathbb{R}^k$ of its vertices in k-dimensional space, we perform the five steps detailed below to obtain $\Delta_{\mathcal{E}}(G)$, a *divergence score* for the embedding. We can apply this algorithm to compare several embeddings $\mathcal{E}_1, \ldots, \mathcal{E}_m$, and select the best one via $\arg\min_{i \in [m]} \Delta_{\mathcal{E}_i}(G)$ (here and later in the paper, we use $[n]$ to denote the set of natural numbers less than or equal to n; that is, $[n] := \{1, \ldots, n\}$). Note that our algorithm is a general framework and some parts have flexibility. We clearly identify these below.

Step 1: Run some stable *graph* clustering algorithm on G to obtain a partition $\mathbf{C}$ of the vertex set V into ℓ communities $C_1, \ldots, C_\ell$.

Note: In our implementation, we used the ensemble clustering algorithm for graphs (ECG) which is based on the Louvain algorithm and the concept of consensus clustering [5], and is shown to have good stability.

Note: In some applications, the desired partition can be provided together with a graph (for example, when nodes contain some natural labelling and so some form of a ground-truth is provided).

Step 2: For each $i \in [\ell]$, let c_i be the proportion of edges of G with both endpoints in C_i. Similarly, for each $1 \le i < j \le \ell$, let $c_{i,j}$ be the proportion of edges of G with one endpoint in C_i and the other one in C_j. Let

$$\bar{\mathbf{c}} = (c_{1,2}, \ldots, c_{1,\ell}, c_{2,3}, \ldots, c_{2,\ell}, \ldots, c_{\ell-1,\ell}) \quad \text{and} \quad \hat{\mathbf{c}} = (c_1, \ldots, c_\ell) \qquad (1)$$

be two vectors with a total of $\binom{\ell}{2} + \ell = \binom{\ell+1}{2}$ entries which together sum to one. These *graph vectors* characterize the partition $\mathbf{C}$ from the perspective of the graph G.

Note: The embedding $\mathcal{E}$ does *not* affect the vectors $\bar{\mathbf{c}}$ and $\hat{\mathbf{c}}$. They are calculated purely based on G and the partition $\mathbf{C}$.

Step 3: For a given parameter $\alpha \in \mathbb{R}_+$ and the same vertex partition $\mathbf{C}$, we consider $\mathcal{G}(\mathbf{w}, \mathcal{E}, \alpha)$, the GCL Model presented in Sect. 3. For each $1 \le i < j \le \ell$, we compute $b_{i,j}$, the expected proportion of edges of $\mathcal{G}(\mathbf{w}, \mathcal{E}, \alpha)$ with one endpoint in C_i and the other one in C_j. Similarly, for each $i \in [\ell]$, let b_i be the expected proportion of edges within C_i. That gives us another two vectors

$$\bar{\mathbf{b}}_{\mathcal{E}}(\alpha) = (b_{1,2}, \ldots, b_{1,\ell}, b_{2,3}, \ldots, b_{2,\ell}, \ldots, b_{\ell-1,\ell})$$
$$\hat{\mathbf{b}}_{\mathcal{E}}(\alpha) = (b_1, \ldots, b_\ell) \qquad (2)$$

with a total of $\binom{\ell+1}{2}$ entries which together sum to one. These *model vectors* characterize the partition $\mathbf{C}$ from the perspective of the embedding $\mathcal{E}$.

Note: The structure of graph G does *not* affect the vectors $\bar{\mathbf{b}}_{\mathcal{E}}(\alpha)$ and $\hat{\mathbf{b}}_{\mathcal{E}}(\alpha)$; only its degree distribution $\mathbf{w}$ and embedding $\mathcal{E}$ are used.

Note: We used the Geometric Chung-Lu Model but the framework is flexible. If, for any reason (perhaps there are some restrictions for the maximum edge length; such restrictions are often present in, for example, wireless networks) it makes more sense to use some other model of random geometric graphs, it can be easily implemented here. If the model is too complicated and computing the expected number of edges between two parts is challenging, then it can be approximated easily via simulations.

Step 4: Compute the distances between the two pairs of vectors, that is, between $\bar{\mathbf{c}}$ and $\bar{\mathbf{b}}_{\mathcal{E}}(\alpha)$, and between $\hat{\mathbf{c}}$ and $\hat{\mathbf{b}}_{\mathcal{E}}(\alpha)$, in order to measure how well the model $\mathcal{G}(\mathbf{w}, \mathcal{E}, \alpha)$ fits the graph G. Let Δ_{α} be a weighted average of the two distances.

Note: We used the well-known and widely used Jensen–Shannon divergence (JSD) to measure the dissimilarity between two probability distributions. The JSD can be viewed as a smoothed version of the Kullback-Leibler divergence. In our implementation, we used simple average, that is,

$$\Delta_{\alpha} = \frac{1}{2} \cdot \left(JSD(\bar{\mathbf{c}}, \bar{\mathbf{b}}(\alpha)) + JSD(\hat{\mathbf{c}}, \hat{\mathbf{b}}(\alpha)) \right). \tag{3}$$

We decided to independently treat internal and external edges to compensate the fact that there are $\binom{\ell}{2}$ coefficients related to external densities whereas only ℓ ones related to internal ones. Depending on the application at hand, other weighted averages can be used if more weight needs to be put on internal or external edges.

Step 5: Select $\hat{\alpha} = \arg\min_{\alpha} \Delta_{\alpha}$, and define the *divergence score* for embedding $\mathcal{E}$ on G as: $\Delta_{\mathcal{E}}(G) = \Delta_{\hat{\alpha}}$.

Note: The parameter α is used to define a distance in the embedding space, as we detail in Sect. 3. In our implementation we simply checked values of α on a grid between 0 and 10. There are clearly better ways to search the space of possible values of α but, since the algorithm worked very fast on our graphs, we did not optimize this part.

In order to compare several embeddings for the same graph G, we repeat steps 3–5 above and compare the divergence scores (the lower the score, the better). Let us stress again that steps 1–2 are done only once, so we use the same partition of the graph into ℓ communities for each embedding. The code can be accessed at the following GitHub repository[1].

2.3   Illustration

We illustrate our framework on the well-known Zachary's Karate Club graph [6]. The parameter $\alpha \geq 0$ in the GCL model controls the distance used in the embedding space. With $\alpha = 0$, the embedding is *not* taken into account and the classic Chung-Lu model is obtained, so only the degree distribution is accounted for. As α gets larger, long edges in the embedding space are penalized more severely. In the left plot of Fig. 1, we show the impact of varying α on the two components of Eq. (3) which respectively consider pairs of vertices that are *internal* (to some

---

[1] https://github.com/ftheberge/Comparing_Graph_Embeddings.

cluster) or *external* (between clusters). Recall that the divergence score for a given embedding is obtained by choosing $\hat{\alpha} = \arg \min_\alpha \Delta_\alpha$. In the right plot of Fig. 1, we used UMAP (Uniform Manifold Approximation and Projection) [7] to show a 2-dimensional projection of the best embedding as obtained by our framework (with node2vec, 64 dimensions and parameters $p = 0.5$ and $q = 1.0$). The vertices are coloured according to the two known communities.

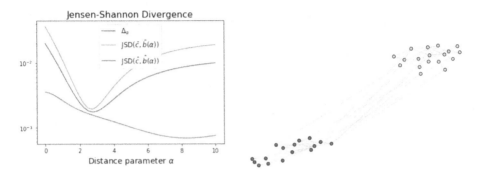

Fig. 1. Zachary's Karate Club Graph. We illustrate the divergence score as a function of α (left) for the best embedding found by our framework (right). The colors represent the two ground-truth communities.

We can use the GCL model to generate edges, as with the standard Chung-Lu model. In Fig. 2, we generate 3 such graphs using the best embedding shown in Fig. 1. The left plot uses $\alpha = 0$, which ignores the embedding and clearly generates too many long edges between the clusters. The center plot uses the optimal value ($\hat{\alpha} = 2.75$ in this case), generating a graph that resembles the true one. The rightmost plot uses the larger value $\alpha = 7$, which penalizes long edges more severely, yielding a graph with less edges between the two communities.

Fig. 2. Zachary's Karate Club Graph. We generate random edges following the Geometric Chung-Lu Model with the same expected degree distribution and with the highest scoring embedding. We look at three cases: $\alpha = 0$ which ignores the embedding (left), $\alpha = 7$ which penalizes long edges too severely (right), and the best $\hat{\alpha} = 2.75$ (center).

3   Geometric Chung-Lu Model

It is known that classical Erdős-Rényi (binomial) random graphs $G(n, p)$ can be generalized to the Chung-Lu model $G(\mathbf{w})$, the random graph with a given expected degree distribution $\mathbf{w} = (w_1, \ldots, w_n)$. It is a classic and well-known model but unfamiliar reader is directed to, for example, [1] or [8]. Since our goal is to compare different embeddings of the same graph, we will generalize the Chung-Lu model further, including geometry coming from the graph embedding. In such models, that are called *spatial models* or *geometric graphs*, vertices are embedded in some metric space and link formation is influenced by the metric distance between vertices. The main principle of spatial models is that vertices that are metrically close are more likely to link to each other. This is a formal expression of the intuitive notion we have about virtual networks: Web links are likely to point to similar pages, people that share similar interests are more likely to become friends on Facebook, and scientific papers mostly refer to papers on similar topics.

In the Geometric Chung-Lu model we are not only given the expected degree distribution of a graph G

$$\mathbf{w} = (w_1, \ldots, w_n) = (\deg_G(v_1), \ldots, \deg_G(v_n))$$

but also an embedding of vertices of G in some k-dimensional space, function $\mathcal{E} : V \to \mathbb{R}^k$. In particular, for each pair of vertices, v_i, v_j, we know the distance between them:

$$d_{i,j} = \text{dist}(\mathcal{E}(v_i), \mathcal{E}(v_j)).$$

It is desired that the probability that vertices v_i and v_j are adjacent to be a function of $d_{i,j}$, that is, to be proportional to $g(d_{i,j})$ for some function g. The function g should be a decreasing function as long edges should occur less frequently than short ones. There are many natural choices such as $g(d) = d^{-\beta}$ for some $\beta \in [0, \infty)$ or $g(d) = \exp(-\gamma d)$ for some $\gamma \in [0, \infty)$. We use the following, normalized function $g : [0, \infty) \to [0, 1]$: for a fixed $\alpha \in [0, \infty)$, let

$$g(d) := \left(1 - \frac{d - d_{\min}}{d_{\max} - d_{\min}}\right)^{\alpha},$$

where

$$d_{\min} = \min\{\text{dist}(\mathcal{E}(v), \mathcal{E}(w)) : v, w \in V, v \neq w\}$$
$$d_{\max} = \max\{\text{dist}(\mathcal{E}(v), \mathcal{E}(w)) : v, w \in V\}$$

are the minimum, and respectively the maximum, distance between vertices in embedding $\mathcal{E}$. One convenient and desired property of this function is that it is invariant with respect to an affine transformation of the distance measure. Clearly, $g(d_{\min}) = 1$ and $g(d_{\max}) = 0$; in the computations, we can use clipping to force $g(d_{\min}) < 1$ and/or $g(d_{\max}) > 0$ if required. Let us also note that if $\alpha = 0$ (that is, $g(d) = 1$ for any $d \in [0, \infty)$ with $g(d_{\max}) = 0^0 = 1$), then we recover the original Chung-Lu model as the pairwise distances are neglected. Moreover, the larger parameter α is, the larger the aversion to long edges is. Since this family of functions (for various values of the parameter α) captures a wide spectrum of behaviours, it should be enough to concentrate on this choice but one can easily experiment with other functions. So, for now we may assume that the only parameter of the model is $\alpha \in [0, \infty)$.

The *Geometric Chung-Lu* (GCL) model is the random graph $G(\mathbf{w}, \mathcal{E}, \alpha)$ on the vertex set $V = \{v_1, \ldots, v_n\}$ in which each pair of vertices v_i, v_j, independently of other pairs, forms an edge with probability $p_{i,j}$, where

$$p_{i,j} = x_i x_j g(d_{i,j})$$

for some carefully tuned weights $x_i \in \mathbb{R}_+$. The weights are selected such that the expected degree of v_i is w_i; that is, for all $i \in [n]$

$$w_i = \sum_{j \in [n], j \neq i} p_{i,j} = x_i \sum_{j \in [n], j \neq i} x_j g(d_{i,j}).$$

Additionally, we set $p_{i,i} = 0$ for $i \in [n]$. Let us mention one technical assumption. It might happen that $p_{i,j}$ is greater than one and so it should really be regarded as the expected number of edges between v_i and v_j; for example, as suggested in the book of Newman [2], one can introduce a Poisson-distributed number of edges with mean $p_{i,j}$ between each pair of vertices v_i, v_j.

In [1], we proved that there exists the unique selection of weights, provided that the maximum degree of G is less than the sum of degrees of all other vertices. Since each connected component of G can be embedded independently, we may assume that G is connected and so the minimum degree of G is at least 1. As a result, this very mild condition is trivially satisfied unless G is a star on n vertices.

It is not clear how to find weights explicitly but they can be easily (and efficiently) approximated numerically to any desired precision, as is discussed in detail in [1].

4   Complexity—Scalable Algorithm

The original algorithm proposed in [1] has a running time that is quadratic as a function of the number of vertices. It was enough to experiment with graphs on a few thousands of vertices to show that the proposed method is feasible and has practical potential (the so-called proof-of-concept). In this section, we improve the complexity and design a scalable algorithm that efficiently evaluates graph embeddings even on millions of vertices.

The main bottleneck of the original algorithm is the process of tuning n weights $x_i \in \mathbb{R}_+$ ($i \in [n]$) in the Geometric Chung-Lu model (Step 3 of the algorithm). This part requires $\Theta(n^2)$ steps and so it is not feasible for large graphs. The other components are much faster with the graph clustering algorithm (Step 1 of the algorithm) being the next computationally intensive part, typically requiring $O(n \ln n)$ steps. We modify our algorithm slightly to obtain a scalable approximation algorithm that can be efficiently run on large networks. Its running time is $O(n \ln n)$ which is practical. Indeed, let us point out that graph embedding algorithms have their own complexity and so our benchmark framework is certainly not a bottleneck of the whole process anymore.

Recall that in Part 3 of the algorithm, for a given parameter $\alpha \in \mathbb{R}_+$ and vertex partition $\mathbf{C}$, we need to compute the expected proportion of edges of $\mathcal{G}(\mathbf{w}, \mathcal{E}, \alpha)$ that are present within partition parts and between them, vectors $\bar{\mathbf{b}}_{\mathcal{E}}(\alpha)$ and $\hat{\mathbf{b}}_{\mathcal{E}}(\alpha)$ defined in (2). The main idea behind our approximation algorithm is quite simple. Our goal is to group together vertices from the same part of $\mathbf{C}$ that are close to each other in the embedded space. Once such refinement of partition $\mathbf{C}$ is generated, we simply replace each group by the corresponding auxiliary vertex that is placed in the (appropriately weighted) center of mass of the group it is associated with. Such auxiliary vertices will be called **landmarks**. Finally, vectors $\bar{\mathbf{b}}_{\mathcal{E}}(\alpha)$ and $\hat{\mathbf{b}}_{\mathcal{E}}(\alpha)$ will be approximated by vectors $\bar{\mathbf{a}}_{\mathcal{E}}(\alpha)$ and $\hat{\mathbf{a}}_{\mathcal{E}}(\alpha)$ in the corresponding auxiliary graph of landmarks. Since we aim for a fast algorithm, the numer of landmarks should be close to $n' = \sqrt{n}$ so that the process of tuning weights can be done in $O(n'^2) = O(n)$ time.

The process of selecting landmarks is discussed in the next subsection but let us mention about one more modification that needs to be done. Our initial Geometric Chung-Lu model produces simple graphs. On the other hand, after merging vertices from one group into the corresponding landmark, we need to control the expected number of edges between these vertices. Hence, we need to generalize our model to include loops which we discuss in the following subsection before we move to the quality and speed comparison.

Generating Landmarks

We start with a partition $\mathbf{C}$ of the vertex set V into ℓ communities $C_1, \ldots, C_\ell$. The number of communities is typically relatively small. In what we write below, our mild assumption is that $\ell < \sqrt{n}$; otherwise, one may simply use the original algorithm or increase the number of landmarks (alternatively, one may insist that the number of initial communities produced by graph clustering algorithm is small). For each part C_i ($i \in [\ell]$) we compute the **weighted center of mass** p_i and the **weighted sum of squared errors (SSE)** e_i, that is,

$$ p_i := \frac{\sum_{j \in C_i} w_j \, \mathcal{E}(v_j)}{\sum_{j \in C_i} w_j} \qquad \text{and} \qquad e_i = \sum_{j \in C_i} w_j \, \text{dist}\big(p_i, \mathcal{E}(v_j)\big)^2 . $$

(Recall that w_j is the degree of vertex v_j and $\mathcal{E}(v_j)$ is its position in the embedded space R^k.) The weighted sum of squared errors is a natural measure of variation within a cluster.

We will refine the partition $\mathbf{C}$ by repeatedly splitting some parts of it with the goal to reach precisely $\sqrt{n}$ parts. However, before we explain which parts will be split, let us concentrate on splitting a given part C_i. The goal is to partition C_i with SSE equal to e_i into two parts with the corresponding SSEs equal to e_i^1 and

e_i^2 in such a way that $\max\{e_i^1, e_i^2\}$ is as small as possible. Finding the best partition is difficult and computationally expensive. However, this can be efficiently well approximated by finding the first principal component in the well-known **weighted Principal Component Analysis (PCA)**. This transformation is defined in such a way that the first principal component has the largest possible weighted variance (that is, accounts for as much of the weighted variability as possible). After projecting all the points from C_i onto this component, we get a total order of these points and one can quickly check which of the natural $|C_i| - 1$ partitions minimizes $\max\{e_i^1, e_i^2\}$. The original part C_i is then replaced with two parts, C_i^1 and C_i^2, with the corresponding centers of mass and SSEs. See Fig. 3 for an illustration of this process.

Splitting C_i into two parts so as to minimize $\max\{e_i^1, e_i^2\}$ can be achieved in $O(|C_i|)$ steps using the bisection search over a projection of data onto the first principal component. Indeed, this is doable because of the following: (1) finding the first principal component and calculating the projection onto it has linear cost, (2) finding the median over some range has a linear cost, (3) when a splitting hyperplane is moved, and as a consequence points between C_i^1 and C_i^2 are moved, then e_i^1 and e_i^2 can be updated using an online algorithm that also has a linear cost, and (4) the number of potentially moved points in the bisection search is halved in each step. (See `split_cluster_rss` function in the reference implementation. In the code we make a significant use of the fact, that the Julia language provides an efficient implementation of views into arrays which allowed us to keep the number of required memory allocations made in the code small.) As a result, since we will be recursively applying splitting until reaching $\sqrt{n}$ parts, similarly as in the case of well-known Quicksort sorting algorithm, the expected total running time of this part of the algorithm is $O(n \ln n)$.

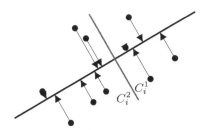

Fig. 3. Splitting C_i using SSE and the first principal component. Dots represent original points, thick black line represents the first principal component, and blue line represents the hyperplane orthogonal to the first principal component. It provides the desired split of C_i into C_i^1 and C_i^2.

Now, we are ready to describe the strategy for selecting parts for splitting. First of all, let us mention that it is not desired to replace the whole original part by one landmark as it may introduce large error. Indeed, the intuition is that replacing many vertices with a landmark might affect the expected number of edges between them but the expected number of edges between vertices that belong to different landmarks (that are often far away from each other) is not affected too much. As a result, in our implementation we insist on splitting each original part into s smaller parts even if the original SSE is small. (s is a parameter of the model that we will discuss soon.) After this initial phase we start splitting parts in a greedy fashion, each time selecting a part that has the largest SSE. The process stops once $n' = \sqrt{n}$ parts are generated.

Let us now briefly discuss the influence of the parameter s. In a typical scenario, even if s is small, each cluster is split many times in the second part of the process where we greedily split clusters with large SSE. In such situations, the value of the parameter s actually does not matter and this is what we observed in our experiments. However, it is theoretically possible that in some rare cases this natural splitting might not happen. As a result, in the implementation we provided, we allow the user to tune parameter s to cover such rare instances.

Now, let us come back to the algorithm. As already mentioned, each part C_i is replaced by its landmark u_i. The position of landmark u_i in the embedded space $\mathbb{R}^k$ coincides with the weighted center of mass of its part, that is, $\mathcal{E}(u_i) = p_i$. Finally, the expected degree of landmark u_i (that we denote as w'_i in order to distinguish it from w_i, the expected degree of vertex w_i) is the sum of the expected degrees of the associated vertices in the original model, that is, $w'_i := \sum_{j \in C_i} w_j$.

Note: We experimented with a number of different strategies for splitting, other than minimizing the maximum SSE, such as balancing sizes of all clusters and balancing diameters of all clusters. Once the objective function is fixed, the algorithm may greedily select the worst cluster (from a given perspective) and then split it appropriately (again, to minimize the objective function). The results were comparable across all strategies.

Including Loops in the Geometric Chung-Lu Model

In order to approximate vectors $\bar{\mathbf{b}}_{\mathcal{E}}(\alpha)$ and $\hat{\mathbf{b}}_{\mathcal{E}}(\alpha)$ from the original model on n vertices, we will use the auxiliary model on $n' = \sqrt{n}$ landmarks. Each landmark u_i is located at $p_i \in \mathbb{R}^k$ (the weighted center of mass of the associated vertices) and has expected degree w_i (the sum of expected degrees of the associated vertices). One can find the pairwise distances between landmarks, and apply the original model $G(\mathbf{w}, \mathcal{E}, \alpha)$ for landmarks to compute the expected number of edges between and within parts, vectors $\bar{\mathbf{a}}_{\mathcal{E}}(\alpha)$ and $\hat{\mathbf{a}}_{\mathcal{E}}(\alpha)$, as an approximation of the original vectors $\bar{\mathbf{b}}_{\mathcal{E}}(\alpha)$ and $\hat{\mathbf{b}}_{\mathcal{E}}(\alpha)$. It is expected that $\bar{\mathbf{a}}_{\mathcal{E}}(\alpha)$ approximates well $\bar{\mathbf{b}}_{\mathcal{E}}(\alpha)$ but, since many vertices ($\sqrt{n}$ on average) are reduced to one landmark, the number of internal edges might be affected, that is, $\hat{\mathbf{a}}_{\mathcal{E}}(\alpha)$ might not be very close to $\hat{\mathbf{b}}_{\mathcal{E}}(\alpha)$.

We partially address this issue by insisting that each original part is split into a number of landmarks. In order to achieve even better approximation we introduce loops in our Geometric Chung-Lu Model. This generalization is straightforward. The *Geometric Chung-Lu* (GCL) model is the random graph

$H(\mathbf{w}, \mathcal{E}, \alpha)$ on the set of landmarks $V = \{u_1, \ldots, u_{n'}\}$ in which each pair of landmarks u_i, u_j, independently of other pairs, forms an edge with probability $p_{i,j}$, where

$$p_{i,j} = x_i x_j g(d_{i,j})$$

for some carefully tuned weights $x_i \in \mathbb{R}_+$. Additionally, for $i \in [n']$, the probability of creating a self loop around landmark u_i is equal to

$$p_{i,i} = x_i^2 g(d_{i,i}), \qquad \text{where} \qquad d_{i,i} = \sqrt{\frac{e_i}{\sum_{j \in C_i} w_j}}.$$

Note that the "distance" $d_{i,i}$ from landmark u_i to itself is an approximation of the unobserved weighted average distance $d_{a,b}$ over all pairs of vertices a and b associated with u_i. The weights are selected such that the expected degree of landmark u_i is w_i'; that is, for all $i \in [n']$

$$w_i' = \sum_{j \in [n']} p_{i,j} = x_i \sum_{j \in [n']} x_j g(d_{i,j}).$$

Since it is an extension to [1], we revisited the proof of the uniqueness of weights in this more general setting (the proof is omitted here due to page limit). We showed that the weights exist and are unique if and only if the following condition is satisfied for more than two landmarks (for completeness, in the full version of the paper we also derived the conditions for the graph on $n' = 2$ landmarks, which are slightly different):

$$d_{t,t} > 0 \vee 2w_t' < \sum_{j \in n'} w_j',$$

where $t = \arg\max_{j \in n'} w_j'$. Finally, let us mention that, as in the case of the original model, standard root-finding algorithms can be used to efficiently find the desired weights.

Quality and Speed Comparison

We start our experiments with the College Football graph and testing the same set of embeddings as in [1]. This well-studied graph with known community structure represents the schedule of United States football games between Division IA colleges during the regular season in Fall 2000 [9]. The data consists of 115 teams (vertices) and 613 games (edges). For each embedding, we compared the original divergence score computed for $n = 115$ vertices with the approximated counterpart computed for $n' = 36$ landmarks. (The value of 36 was selected rather arbitrarily; the graph is tiny so any number of landmarks seems reasonable for this illustration purpose.) Each of the 12 clusters were forced to split once before greedy strategy was applied. The graph presented in Fig. 4 shows very high correlation between the two measures which indicates that the approximation algorithm preforms well, as expected.

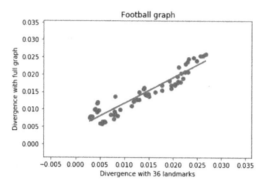

Fig. 4. College Football Graph exhibits high correlation between the original divergence score and its approximated counterpart.

We compared the two sets of divergence scores for all embeddings, the first set based on the original algorithm and the second one based on the approximated version. The two sets of scores (as well as their rankings) are highly correlated as indicated by the following two measures of similarity: Pearson's correlation of 0.941 for the divergence scores and Kendall-tau of 0.802 for the rankings. Having said that, the rankings that we obtained are not identical. In Fig. 5, we show the best and worst scoring embeddings for the approximated divergence score based on landmarks. The conclusion is the same as for the original algorithm: embeddings that score high are of good quality wheres the ones that score low are of poor quality.

Fig. 5. The College Football Graph. We show the best (left) and the worst (right) scoring embedding based on the approximated algorithm with landmarks.

Our next experiment is with Email-Eu-core Network on 986 vertices. This network was generated using email data from a large European research institution and is available as one of the SNAP Datasets [10]. As before, we tested all available embeddings. Clearly, our approximation algorithm provides a trade-off between the speed and the accuracy of the obtained approximation—the more landmarks we use, the better approximation we get but the algorithm gets slower. The goal of this experiment is to investigate how sensitive the approximation is as a function of the number landmarks. We compare the Pearson's correlation between the two sets of divergence scores of all embeddings, the first one computed for the original graph on $n = 986$ vertices, and the second one computed for the approximated variant on n' landmarks with $n' \geq 25$—see Fig. 6. As expected, there is a high correlation between the two sets with a satisfying outcomes already around $\sqrt{n}$ landmarks.

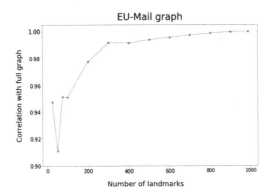

Fig. 6. Pearson's correlation between the divergence scores computed for the original graph on $n = 986$ vertices and the ones computed for the approximated variant on $n' \geq 25$ landmarks.

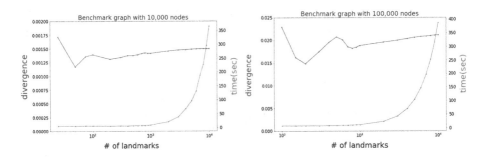

Fig. 7. Comparing quality and speed for ABCD graphs. We compare the approximated divergence score and the time required to compute it as a function of the number of landmarks.

In order to see how the approximated algorithm behaves on large graphs, our last experiments are concerned with relatively large instances of ABCD graphs, on $n = 10{,}000$ vertices and on $n = 100{,}000$ vertices. The Artificial Benchmark for Community Detection (ABCD graph) [11] is a random graph model with community structure and power-law distribution for both degrees and community sizes (a new model that is an attempt to solve some of the problems of the standard method for generating artificial networks, the LFR graph generator [12]). Whereas the smaller graph can be easily tested by the original algorithm, dealing with the larger graph seems impractical (we only tested it for $n' \leq 10{,}000 < n = 100{,}000$ landmarks). On the other hand, the approximated algorithm easily deals with graphs of that size.

For each graph, we tested the embedding obtained by 8-dim node2vec algorithm (time required to generate embedding for the small graph was roughly 32 s wheres the large graph required 6 min and 20 s to be processed). The results are presented in Fig. 7. For each number of landmarks n', we plot the approximated divergence score as well as the time required to compute it on 2.2 GHz Intel Xeon E5 processor. We clearly see the trade-off between the accuracy and the speed of the algorithm with the "sweet spot" around $n' \approx \sqrt{n}$ where the approximated divergence score is very close to the original divergence score whereas the algorithm is still extremely fast. In order to reach that conclusion, we tested the algorithm for the values of n' up to $n' = 10^4 = (10^5)^{4/5} = n^{4/5}$, much larger than $n' = \sqrt{n}$. As a result, in practice, one can easily deal with graphs or order $n = (10^4)^2 = 10^8$.

5   Future Directions

In this paper, our aim was to introduce a scalable general framework for evaluating embeddings. This exploratory research showed that our divergence score is a very promising distinguisher. The next natural step is to do extensive experiments of various embedding algorithms on large real-world datasets in order to evaluate and compare them.

A further extension of this work could be made to weighted graphs or hypergraphs that are generalizations of graphs in which a single (hyper)edge can connect any number of vertices. Hypergraphs are often more suitable and useful for representing and modelling many important networks and processes. We are interested in generalizing classic notions to hypergraphs, such as clustering via modularity [13], as well as developing new algorithms to apply to them [14]. Hence, a natural line of development of the proposed embedding comparison framework is to generalize it to allow for evaluation of embeddings of hypergraphs.

As a side effect of our research on evaluating graph embeddings, we have introduced the Geometric Chung-Lu model that is interesting on its own right and potentially applicable in other problems. As it is not the main focus of this paper, we did not analyze its graph-theoretic properties in detail. It remains as a subject for further research.

References

1. Kamiński, B., Prałat, P., Théberge, F.: An unsupervised framework for comparing graph embeddings. J. Complex Networks, in press. 27 p
2. Newman, M.: Networks: An Introduction. Oxford University Press, Oxford (2010)

3. Bianconi, G.: Interdisciplinary and physics challenges of network theory. EPL **111**(5), 56001 (2015)
4. Janssen, J.: Spatial models for virtual networks. In: Ferreira, F., Löwe, B., Mayordomo, E., Mendes Gomes, L. (eds.) CiE 2010. LNCS, vol. 6158, pp. 201–210. Springer, Heidelberg (2010). https://doi.org/10.1007/978-3-642-13962-8_23
5. Poulin, V., Théberge, F.: Ensemble clustering for graphs. In: Aiello, L.M., Cherifi, C., Cherifi, H., Lambiotte, R., Lió, P., Rocha, L.M. (eds.) COMPLEX NETWORKS 2018. SCI, vol. 812, pp. 231–243. Springer, Cham (2019). https://doi.org/10.1007/978-3-030-05411-3_19
6. Zachary, W.W.: An information flow model for conflict and fission in small groups. J. Anthropol. Res. **33**, 452–473 (1977)
7. McInnes, L., Healy, J., Melville, J.: UMAP: uniform manifold approximation and projection for dimension reduction. pre-print arXiv:1802.03426 (2018)
8. Chung, F.R.K., Lu, L.: Complex Graphs and Networks. American Mathematical Society, Boston (2006)
9. Girvan, M., Newman, M.E.: Community structure in social and biological networks. Proc. Natl. Acad. Sci. **99**, 7821–7826 (2002)
10. Leskovec, J., Krevl, A.: SNAP datasets: Stanford large network dataset collection. http://snap.stanford.edu/data
11. Kamiński, B., Prałat, P., Théberge, F.: Artificial benchmark for community detection (ABCD) – fast random graph model with community structure, pre-print arXiv:2002.00843 (2020)
12. Lancichinetti, A., Fortunato, S., Radicchi, F.: Benchmark graphs for testing community detection algorithms. Phys. Rev. E **78**(4), 046110 (2008)
13. Kamiński, B., Poulin, V., Prałat, P., Szufel, P., Théberge, F.: Clustering via Hypergraph Modularity. PLoS ONE **14**(11), e0224307 (2019)
14. Antelmi, A., et al.: Analyzing, exploring, and visualizing complex networks via hypergraphs using SimpleHypergraphs.jl. Internet Math. (2020). 32 p

Assortativity and Bidegree Distributions on Bernoulli Random Graph Superpositions

Mindaugas Bloznelis[1], Joona Karjalainen[2], and Lasse Leskelä[2(✉)]

[1] Institute of Informatics, Vilnius University,
Naugarduko 24, 03225 Vilnius, Lithuania
[2] Department of Mathematics and Systems Analysis, School of Science,
Aalto University, Otakaari 1, 02015 Espoo, Finland
lasse.leskela@aalto.fi

Abstract. A probabilistic generative network model with n nodes and m overlapping layers is obtained as a superposition of m mutually independent Bernoulli random graphs of varying size and strength. When n and m are large and of the same order of magnitude, the model admits a sparse limiting regime with a tunable power-law degree distribution and nonvanishing clustering coefficient. This article presents an asymptotic formula for the joint degree distribution of adjacent nodes. This yields a simple analytical formula for the model assortativity, and opens up ways to analyze rank correlation coefficients suitable for random graphs with heavy-tailed degree distributions.

Keywords: Joint degree distribution · Bidegree distribution · Degree–degree distribution · Empirical degree distribution · Degree correlation · Transitivity · Statistical network model · Erdős–Rényi graph · Random intersection graph

1 Introduction

Overview and Objectives. Questions in technology, life sciences, and economics are often related to large systems of nodes connected via pairwise interactions which involve uncertainty due to unpredictable node behavior and missing data. Such uncertainties have been mathematically modeled and analyzed using random graph models of various complexity, including classical independently linked and uniform random graphs [17], stochastic block models and inhomogeneous Bernoulli graphs [1,12,20], random graphs with given degree distributions [13,32], and generative models involving preferential attachment and rewiring mechanisms [3,35]. While succeeding to obtain a good fit to degree distributions, most earlier models fail to capture second-order effects related to clustering and transitivity. Random intersection graphs [6,7,15,18,23,25], spatial preferential attachment models [2,21], and hyperbolic random geometric graphs [11,26,27]

© Springer Nature Switzerland AG 2020
B. Kamiński et al. (Eds.): WAW 2020, LNCS 12091, pp. 68–81, 2020.
https://doi.org/10.1007/978-3-030-48478-1_5

have been successful in extending the analysis to sparse graph models with tunable global clustering coefficient. Despite remarkable methodological advances obtained in the aforementioned articles and related literature, most models of sparse random graphs still appear somewhat rigid in what comes to modeling finer second-order properties, such as correlations of the degrees of adjacent nodes [34] and degree-dependent clustering coefficients [5,42].

Main Contributions. This article discusses a mathematical network model recently introduced in [10] which is motivated by the structure of social networks composed of a large number of overlapping communities [14]. The model is generated as a superposition of mutually independent Bernoulli random graphs $G_1, \ldots, G_m$ of variable size (number of nodes) and strength (link probability), which can be interpreted as *layers* or *communities*. The node sets of the layers are *random* subsets of the underlying population of n nodes. A key feature of the model is that the layer sizes and layer strengths are assumed to be correlated, which allows for example to model social networks with tunable frequencies of strong small communities and weak large communities. The main contribution of this article is a rigorous mathematical analysis (Theorem 1) of the bidegree distribution (joint degree distribution of adjacent nodes) of the model in a limiting regime where the number of nodes n and the number of layers m are large and of the same order of magnitude. The bidegree distribution yields compact mathematical formulas for model assortativity (Theorem 2) and rank correlations (Theorem 3) of the adjacent node degrees. The latter theorem is suitable for modeling dependencies in heavy-tailed models with degrees having unbounded second moments.

Related Work. Degree distributions, clustering, and percolation analysis of the model is presented in [10]. An analogous model where the node sets of the layers are deterministic has been studied in [44] in the context of overlapping community detection. Clustering coefficients and small subgraph frequencies for a special case with constant layer strengths have been analyzed in [19,23,24,36]. In the special case with unit layer strengths, the layers become cliques and the model reduces to the passive random intersection graph introduced in [18], with degree and clustering properties analyzed in [7,30]. A network model with similar features has been recently presented in [38]. Assortativity and bidegree distributions have earlier been analyzed in the context of random intersection graph models [8,9], inhomogeneous Bernoulli graphs and their extensions [12,31,37], preferential attachment models [28,39], and configuration models in [39–41]. Extremal properties of bidegree correlations in general graphs have been reported in [16,39].

1.1 Notations

Sets and Numbers. The cardinality of a set A is denoted $|A|$. Ordered pairs are denoted by (i,j), and unordered pairs by $ij = \{i,j\}$. Here $1(A)$ is defined to be one when statement A is true, and zero otherwise. We denote

$[n] = \{1, \ldots, n\}$ and $\mathbb{Z}_+ = \{0, 1, 2, \ldots\}$. The falling factorial is denoted $(x)_r = x(x-1)\cdots(x-r+1)$.

Graphs. A graph is a pair $G = (V(G), E(G))$ where $V(G)$ is a set of elements called nodes, and $E(G)$ is a collection of unordered node pairs. Nodes i and j are called adjacent if $ij \in E(G)$. The set of nodes adjacent to i is denoted $N_G(i) = \{j \in V(G) : ij \in E(G)\}$. The degree of i is denoted $\deg_G(i) = |N_G(i)|$.

Probability. For probability measures on countable spaces we denote $f(x) = f(\{x\})$ and $\int \phi\,df = \sum \phi(x)f(x)$. The Dirac measure at x is denoted by δ_x. The binomial distribution is denoted by $\mathrm{Bin}(x, y)(s) = \binom{x}{s}(1-y)^{x-s}y^s$, and the Poisson distribution by $\mathrm{Poi}(\lambda)(s) = e^{-\lambda}\frac{\lambda^s}{s!}$. The product and the convolution of probability measures f and g are denoted by $f \otimes g$ and $f * g$, respectively.

2   Assortativity and Bidegree Distributions

2.1   Empirical Quantities

Let G be a graph with a finite node set and a nonempty link set. Here G is viewed as a nonrandom graph or a fixed sample of a random graph. The (empirical) *degree distribution* of G is a probability measure on $\mathbb{Z}_+$ defined by

$$f_G(s) = \frac{1}{|V(G)|} \sum_{i \in V(G)} 1(\deg_G(i) = s),$$

and represents the probability distribution of random variable $\deg_G(I)$ where I is a random variable obtained by sampling a node uniformly at random. The (empirical) *bidegree distribution* of G with a nonempty link set is a probability measure on $\mathbb{Z}_+^2$ defined by

$$f_G^{(2)}(s, t) = \frac{1}{2|E(G)|} \sum_{(i,j):\{i,j\}\in E(G)} 1(\deg_G(i) = s, \deg_G(j) = t).$$

This is the joint probability distribution of the pair $(\deg_G(I), \deg_G(J))$ obtained by sampling (I, J) uniformly at random from the set of all ordered node pairs adjacent in G. Both marginals of the bidegree distribution are equal to the size-biased degree distribution $f_G^*(s) = \frac{sf_G(s)}{\sum_t tf_G(t)}$. The Pearson correlation coefficient of the bidegree distribution is called the (empirical) *assortativity* of graph G and can be written as

$$\mathrm{Cor}_G(\deg_G(I), \deg_G(J)) = \frac{\sum_{s,t} st f_G^{(2)}(s, t) - (\sum_s s f_G^*(s))^2}{\sum_s s^2 f_G^*(s) - (\sum_s s f_G^*(s))^2}.$$

2.2   Model Quantities

Let G be a random graph such that $V(G)$ is nonrandom and finite, and $E(G)$ is nonempty with positive probability. The *model degree distribution* of G is defined by

$$f(s) \;=\; \mathbb{P}\big(\deg_G(I) = s\big), \tag{1}$$

where I is a random node in $V(G)$, selected uniformly at random and independently of $E(G)$. The *model bidegree distribution* is defined by

$$f_2(s,t) \;=\; \mathbb{P}\Big(\deg_G(I) = s,\, \deg_G(J) = t \,\big|\, IJ \in E(G)\Big), \tag{2}$$

where (I, J) is an ordered pair of distinct nodes of $V(G)$, selected uniformly at random and independently of $E(G)$. By simple computations one may verify that $f_2(t, s) = f_2(s, t)$, and that both marginals of the model bidegree distribution are equal to the size-biased model degree distribution $f^*(s) = \frac{sf(s)}{\sum_t tf(t)}$. The Pearson correlation coefficient of the model bidegree distribution is called the *model assortativity*, and can be written as

$$\mathrm{Cor}^*(D_I, D_J) \;=\; \frac{\mathbb{E}^* D_I D_J - (\mathbb{E}^* D_I)^2}{\mathbb{E}^* D_I^2 - (\mathbb{E}^* D_I)^2}, \tag{3}$$

where $D_I = \deg_G(I)$ and $D_J = \deg_G(J)$ for (I, J) as above, and $\mathbb{E}^*$ refers to conditional expectation given $\{IJ \in E(G)\}$.

The random graph model is called *exchangeable* if its law is invariant to node permutations. In this case the model degree distribution can be written as in (1) but with I replaced by an arbitrary node i. Similarly, formulas (2)–(3) remain valid with (I, J) replaced by an arbitrary pair (i, j) of distinct nodes.

3   Random Graph Superposition Model

A multilayer network model with n nodes and m layers is defined by a list

$$\Big((G_{n,1}, X_{n,1}, Y_{n,1}), \ldots, (G_{n,m}, X_{n,m}, Y_{n,m})\Big)$$

of mutually independent random variables with values in $\mathcal{G}_n \times \{0, \ldots, n\} \times [0, 1]$, where $\mathcal{G}_n$ is the set of undirected graphs with node set contained in $\{1, \ldots, n\}$. We assume that conditionally on $(X_{n,k}, Y_{n,k})$, the probability distribution of $V(G_{n,k})$ is uniform on the subsets of $\{1, \ldots, n\}$ of size $X_{n,k}$, and conditionally on $(V(G_{n,k}), X_{n,k}, Y_{n,k})$, the probability distribution of $E(G_{n,k})$ is such that each node pair of $V(G_{n,k})$ is linked with probability $Y_{n,k}$, independently of other node pairs. The variables $X_{n,k}, Y_{n,k}$ are called the *size* and *strength* of layer k, respectively. Aggregation of the layers produces an overlay random graph G_n defined by

$$V(G_n) \;=\; \{1, \ldots, n\}$$

and

$$E(G_n) = E(G_{n,1}) \cup \cdots \cup E(G_{n,m}).$$

We obtain a rich class of generative probabilistic models when we assume that the layer types $(X_{n,1}, Y_{n,1}), \ldots, (X_{n,m}, Y_{n,m})$ are mutually independent and distributed according to a probability measure $P^{(n)}$ on $\{0, \ldots, n\} \times [0, 1]$.

A large network is modeled as a sequence of network models of the above type indexed by the number of nodes $n = 1, 2, \ldots$ so that the number of layers $m = m_n$ tends to infinity as $n \to \infty$. To obtain tractable limiting formulas with rich expressive power, we shall focus on a sparse parameter regime where there exists a probability measure P on $\{0, 1, \ldots\} \times [0, 1]$ which approximates the layer type distribution according to $P^{(n)} \to P$ weakly, together with the convergence of suitable cross moments $P_{rs}^{(n)} \to P_{rs}$, where we use the shorthand notations

$$P_{rs}^{(n)} = \mathbb{E}\Big((X_{n,k})_r Y_{n,k}^s\Big), \qquad P_{rs} = \mathbb{E}\Big((X)_r Y^s\Big),$$

with $(x)_r = x(x-1)\cdots(x-r+1)$, and (X, Y) being a generic P-distributed random variable.

Sparse network models with a finite nonzero average degree are obtained when the number of layers is of the same order as the number of nodes. When $\frac{m}{n} \to \mu \in (0, \infty)$, $P^{(n)} \to P$ weakly, and $P_{10}^{(n)} \to P_{10} \in (0, \infty)$, then the model degree distribution of G_n converges weakly [10] to a compound Poisson distribution

$$f = \mathrm{CPoi}(\lambda, g) \tag{4}$$

with rate parameter $\lambda = \mu P_{10}$ and increment distribution

$$g(s) = \int \mathrm{Bin}(x-1, y)(s) \frac{x P(dx, dy)}{P_{10}}, \qquad s \in \mathbb{Z}_+. \tag{5}$$

The limiting model degree distribution f can be represented as the law of $D = \sum_{k=1}^{\Lambda} D_k$, where $\Lambda, D_1, D_2, \ldots$ are mutually independent random integers and such that $\mathrm{Law}(\Lambda) = \mathrm{Poi}(\lambda)$ and $\mathrm{Law}(D_k) = g$.

4   Main Results

4.1   Bidegree Distribution

The result below characterizes the limiting bidegree distribution in the random Bernoulli graph superposition model. The limiting bidegree distribution can be represented as the joint law of random variables

$$(D_1^*, D_2^*) = (1 + D_1 + D_1', \, 1 + D_2 + D_2'), \tag{6}$$

where D_1, D_2, and (D_1', D_2', X', Y') are mutually independent and such that D_1 and D_2 follow the limiting degree distribution f defined by (4), D_1' and D_2' are conditionally independent and $\mathrm{Bin}(X' - 2, Y')$-distributed given (X', Y'), and the joint distribution of $(X', Y') \in \mathbb{Z}_+ \times [0, 1]$ equals $\frac{(x)_2 y \, P(dx, dy)}{P_{21}}$. Here X' and Y' represent the size and strength of a random layer which links a particular node pair $\{i, j\}$, and D_1' and D_2' represent the number of additional neighbors of i and j inside the common layer. The joint distribution of (D_1^*, D_2^*) defined by (6) can be written as

$$f_2 = \delta_{(1,1)} * (f \otimes f) * f_2' \tag{7}$$

where $*$ refers to the convolution of probability measures on $\mathbb{Z}_+^2$, and f_2' is a probability measure on $\mathbb{Z}_+^2$ defined by

$$f_2'(s, t) = \int_{\mathbb{Z}_+ \times [0,1]} \mathrm{Bin}(x - 2, y)(s) \, \mathrm{Bin}(x - 2, y)(t) \frac{(x)_2 y \, P(dx, dy)}{P_{21}}. \tag{8}$$

Theorem 1. *Denote by $f_{2,n}$ the bidegree distribution of the n-th model G_n. Assume that $\frac{m}{n} \to \mu \in (0, \infty)$ and $P^{(n)} \to P$ weakly for some probability measure P on $\mathbb{Z}_+ \times [0, 1]$ such that $P_{21} > 0$.*

(i) If $P_{20}^{(n)} \to P_{20} < \infty$, then $f_{2,n} \to f_2$ weakly, where the limit is defined by (7).
(ii) If in addition, $P_{rs}^{(n)} \to P_{rs} < \infty$ for $rs = 32, 43$, then $\int \phi \, df_{2,n} \to \int \phi \, df_2$ for all $\phi : \mathbb{Z}_+^2 \to \mathbb{R}$ such that $|\phi(x, y)| \le c(1 + x^2 + y^2)$ for some constant $c < \infty$ (convergence in the Wasserstein-2 metric [43, Theorem 6.9]).

4.2 Assortativity

The following result provides a formula of the limiting model assortativity which is well defined when the limiting degree distribution has a finite third moment. In the special case with unit strengths, this formula yields the corresponding result for passive random intersection graphs given in [9, Theorem 3.1]. Using a well-chosen coupling of P-distributed random vectors (details in an extended version) it is possible to verify that

$$P_{21}(P_{43} + P_{32}) - P_{32}^2 \ge P_{21}(P_{43} + P_{33}) - P_{32}^2 \ge 0,$$

which implies that the limiting model assortativity below is always nonnegative.

Theorem 2. *Assume that $\frac{m}{n} \to \mu \in (0, \infty)$, and that $P_{rs}^{(n)} \to P_{rs} < \infty$ for $rs = 20, 32, 43$, for some probability measure P on $\mathbb{Z}_+ \times [0, 1]$ such that $P_{21} > 0$. Then the model assortativity is approximated by*

$$\mathrm{Cor}^*(D_I, D_J) \to \frac{P_{21}(P_{43} + P_{33}) - P_{32}^2}{P_{21}(P_{43} + P_{32}) - P_{32}^2 + \mu P_{21}^2(P_{21} + P_{32})}.$$

4.3 Rank Correlations

Assortativity modeled using Pearson's correlation of the bidegree distribution is ill-behaved for graph models where the limiting degree distribution has an infinite third moment [39]. In such cases, rank correlation coefficients provide a robust alternative [39–41]. For a probability measure f on $\mathbb{R}^2$ with nondegenerate marginals, Kendall's rank correlation [29,33] is defined by

$$\rho_{\text{Ken}}(f) = \text{Cor}\left(\text{sgn}(X_1 - Y_1), \text{sgn}(X_2 - Y_2)\right)$$

where $\text{sgn}(x) = 1(x > 0) - 1(x < 0)$, and (X_1, X_2) and (Y_1, Y_2) are mutually independent and f-distributed. Spearman's rank correlation is defined as

$$\rho_{\text{Spe}}(f) = \text{Cor}\left(r_1(X_1), r_2(X_2)\right),$$

where (X_1, X_2) is f-distributed and $r_i(x) = \frac{1}{2}(f^{(i)}(-\infty, x) + f^{(i)}(-\infty, x])$ with $f^{(i)}$ denoting the i-th marginal distribution of f. There are several alternative definitions for Spearman's rank correlation corresponding to different tie-breaking conventions [4]. The above definition agrees with the commonly used mid-rank convention [33, Theorems 14 and 15].

Theorem 3. *Assume that $\frac{m}{n} \to \mu \in (0, \infty)$, and $P^{(n)} \to P$ weakly with $P_{20}^{(n)} \to P_{20}$, where $0 < P_{21} \le P_{20} < \infty$. Spearman's and Kendall's rank correlation coefficients of the n-th model are then approximated by*

$$\rho_{\text{Ken}}(f_{2,n}) \to \rho_{\text{Ken}}(f_2) \qquad \text{and} \qquad \rho_{\text{Spe}}(f_{2,n}) \to \rho_{\text{Spe}}(f_2),$$

where the limiting bidegree distribution f_2 is defined by (7).

5   Discussion

This article described degree correlations in a sparse network model introduced in [10], constructed by a natural superposition mechanism with overlapping layers. The main contribution is a compact explicit description of the limiting model bidegree distribution (Theorem 1), fully characterized in terms of the limiting joint distribution P of layer sizes and layer strengths, and the limiting ratio μ of the number of layers and the number of nodes. Some remarks deserve further attention.

(i) The model bidegree distribution differs from the empirical bidegree distribution computed from a fixed random graph sample. Several earlier works [39–41] have focused on the convergence in probability of the latter distribution. Based on analogous studies on ergodic properties of clustering coefficient [23,24], we expect that both distributions converge to the same limit under mild regularity assumptions.

(ii) The freedom to tune the limiting joint distribution P of layer sizes and layer strengths yields a rich class of network models. As a concrete example, assume that the layer strength is a deterministic function of layer size such $Y = q(X)$. If layer sizes follow an approximate power law $\mathbb{P}(X = x) \propto x^{-\alpha}$ with $\alpha > 2$, and $q(x) \propto x^{-\beta}$ where $\beta \in [0, 1)$ is such that $\alpha + \beta > 3$, then the limiting degree distribution follows a power law [10] such that $\mathbb{P}(D_1 = t) \propto t^{-\delta}$ with $\delta = 1 + \frac{\alpha-2}{1-\beta}$. (The same functional form of layer strengths has been also investigated in [44] for deterministic layer node sets.) Because the marginals of the limiting bidegree distribution are size-biased versions of the degree distribution, it follows that the marginals of f_2 are power laws with density exponent $\delta - 1 = \frac{\alpha-2}{1-\beta}$. The dependence structure of the power-law random variables D_1^* and D_2^* is implicitly captured by (6). Characterizing how the dependence structure behaves as a function of the power-law exponents is an interesting problem to be considered elsewhere.

(iii) Fitting the model to real data sets is a problem of future research. A fully nonparametric approach to estimating P appears hard if not impossible, even though currently there are no (positive or negative) theoretical results regarding model identifiability. An alternative approach is to restrict to models where $P = P_\theta$ is parametrized by a small-dimensional parameter θ, and develop estimators of θ using empirical small subgraph counts. Recent work in this direction includes [19, 23, 24] for models with constant layer strength. Model fitting with deterministic (unknown) layer node sets has been studied in [44].

6   Proofs

6.1   Correlation of the Limiting Bidegree Distribution

Let us analyze the Pearson correlation coefficient $\mathrm{Cor}(D_1^*, D_2^*)$ of the limiting bidegree distribution in Theorem 1.

Proposition 1. *For any $\mu \in (0, \infty)$ and any probability measure P on $\mathbb{Z}_+ \times [0,1]$ such that $0 < P_{10}, P_{21} < \infty$ and $P_{32}, P_{43} < \infty$, the random variables (D_1^*, D_2^*) in (6) satisfy*

$$\mathrm{Cor}(D_1^*, D_2^*) = \frac{P_{21}(P_{43} + P_{33}) - P_{32}^2}{P_{21}(P_{43} + P_{32}) - P_{32}^2 + \mu P_{21}^2(P_{21} + P_{32})}.$$

Proof. If B is a $\mathrm{Bin}(x - 2, y)$-distributed random variable, then $\mathbb{E}B = (x - 2)y$ and $\mathbb{E}(B)_2 = (x - 2)_2 y^2$, from which we conclude that $\mathbb{E}B^2 = \mathbb{E}(B)_2 + \mathbb{E}B = (x - 2)_2 y^2 + (x - 2)y$. Because $(x - 2)(x)_2 = (x)_3$, it follows that

$$\mathbb{E}D_1' = \int (x - 2)y \frac{(x)_2 y P(dx, dy)}{P_{21}} = \frac{P_{32}}{P_{21}}.$$

Further, by noting that $(x - 2)_2(x)_2 = (x)_4$, we see that

$$\mathbb{E}(D_1')^2 = \int \left((x - 2)_2 y^2 + (x - 2)y \right) \frac{(x)_2 y P(dx, dy)}{P_{21}} = \frac{P_{43} + P_{32}}{P_{21}}.$$

Hence D_1' has a finite second moment, and variance equal to

$$\mathrm{Var}(D_1') = \frac{P_{43} + P_{32}}{P_{21}} - \left(\frac{P_{32}}{P_{21}} \right)^2. \tag{9}$$

Similarly, the conditional independence of D_1' and D_2', together with the formula $(x-2)^2(x)_2 = (x-2)(x)_3 = (x)_4 + (x)_3$, implies that

$$\mathbb{E}D_1'D_2' = \int ((x-2)y)^2 \frac{(x)_2 y P(dx, dy)}{P_{21}} = \frac{P_{43} + P_{33}}{P_{21}},$$

and hence, noting that D_1' and D_2' identically distributed,

$$\mathrm{Cov}(D_1', D_2') = \frac{P_{43} + P_{33}}{P_{21}} - \left(\frac{P_{32}}{P_{21}}\right)^2. \tag{10}$$

Recall next that D_1 follows the compound Poisson distribution $f = \mathrm{CPoi}(\lambda, g)$. A simple computation confirms that the variance of g in (5) equals $\frac{P_{32} + P_{21}}{P_{10}}$. Hence it follows (using basic properties of compound Poisson distributions) that D_1 has a finite second moment with

$$\mathrm{Var}(D_1) = \lambda \frac{P_{32} + P_{21}}{P_{10}}. \tag{11}$$

The mutual independence of D_1, D_2, and (D_1', D_2') implies that $\mathrm{Cov}(D_1^*, D_2^*) = \mathrm{Cov}(D_1', D_2')$ and $\mathrm{Var}(D_1^*) = \mathrm{Var}(D_1) + \mathrm{Var}(D_1')$, so that

$$\mathrm{Cor}(D_1^*, D_2^*) = \frac{\mathrm{Cov}(D_1', D_2')}{\mathrm{Var}(D_1) + \mathrm{Var}(D_1')}. \tag{12}$$

By plugging (9)–(11) into (12), we conclude that

$$\mathrm{Cor}(D_1^*, D_2^*) = \frac{\frac{P_{43} + P_{33}}{P_{21}} - \left(\frac{P_{32}}{P_{21}}\right)^2}{\frac{P_{43} + P_{32}}{P_{21}} - \left(\frac{P_{32}}{P_{21}}\right)^2 + \lambda \frac{P_{32} + P_{21}}{P_{10}}}.$$

By recalling that $\lambda = \mu P_{10}$, the claim follows. $\square$

6.2   Proof Outline of Theorem 1:(i)

Denote the bidegree distribution of the n-th model by

$$f_{2,n}(s, t) = \mathbb{P}\Big(\deg_{G_n}(1) = s, \deg_{G_n}(2) = t \mid 12 \in E(G_n) \Big).$$

For $A \subset [m]$, denote by $G_{n,A}$ the graph with $V(G_{n,A}) = [n]$ and $E(G_{n,A}) = \cup_{k \in A} E(G_{n,k})$. We abbreviate $D_i = \deg_{G_n}(i)$, and we note that for any k,

$$D_i = D_{i,k} + \tilde{D}_{i,k} - \hat{D}_{i,k},$$

where

$$D_{i,k} = \deg_{G_{n,k}}(i), \quad \tilde{D}_{i,k} = \deg_{G_{n,[m]\setminus\{k\}}}(i), \quad \hat{D}_{i,k} = \deg_{G_{n,k} \cap G_{n,[m]\setminus\{k\}}}(i).$$

Also denote $\mathcal{E}_k = \{12 \in E(G_{n,k})\}$ and

$$
\begin{aligned}
f_n(s) &= \mathbb{P}(D_i = s), \\
\tilde{f}_{2,n}(s,t) &= \mathbb{P}(\tilde{D}_{1,k} = s, \tilde{D}_{2,k} = t), \\
f'_{2,n}(s,t) &= \mathbb{P}(D_{1,k} = s, D_{2,k} = t \mid \mathcal{E}_k).
\end{aligned}
$$

The proof is based on approximating (details to appear in an extended version):

$$
\begin{aligned}
&f_{2,n}(s,t)\, \mathbb{P}(12 \in E(G_n)) \\
&= \mathbb{P}(D_1 = s, D_2 = t, \cup_k \mathcal{E}_k) \\
&\approx \sum_k \mathbb{P}(D_1 = s, D_2 = t, \mathcal{E}_k) \\
&= \sum_k \mathbb{P}(D_{1,k} + \tilde{D}_{1,k} - \hat{D}_{1,k} = s,\ D_{2,k} + \tilde{D}_{2,k} - \hat{D}_{2,k} = t,\ \mathcal{E}_k) \\
&\approx \sum_k \mathbb{P}(D_{1,k} + \tilde{D}_{1,k} = s,\ D_{2,k} + \tilde{D}_{2,k} = t,\ \mathcal{E}_k) \\
&= \sum_k \sum_{s_1 \le s}\sum_{t_1 \le t} \tilde{f}_{2,n}(s_1,t_1)\, \mathbb{P}(D_{1,k} = s - s_1, D_{2,k} = t - t_1, \mathcal{E}_k) \\
&= \left(\sum_k \mathbb{P}(\mathcal{E}_k) \right) \sum_{s_1 \le s}\sum_{t_1 \le t} \tilde{f}_{2,n}(s_1,t_1) f'_{2,n}(s - s_1, t - t_1) \\
&\approx \mathbb{P}(12 \in E(G_n)) \sum_{s_1 \le s}\sum_{t_1 \le t} \tilde{f}_{2,n}(s_1,t_1)\, f'_{2,n}(s - s_1, t - t_1) \\
&\approx \mathbb{P}(12 \in E(G_n)) \sum_{s_1 \le s}\sum_{t_1 \le t} f_n(s_1) f_n(t_1)\, f'_{2,n}(s - s_1, t - t_1).
\end{aligned}
$$

As a consequence,

$$
|f_{2,n}(s,t) - ((f_n \otimes f_n) * f'_{2,n})(s,t)| \ \to\ 0 \tag{13}
$$

for any $s,t \in \mathbb{Z}_+$, with $*$ denoting the convolution of probability measures on the additive group $\mathbb{Z}^2$. Next, we know that $f_n \to f$ weakly where f is the limiting model degree distribution in (4). Therefore, $f_n \otimes f_n \to f \otimes f$ weakly as probability measures on $\mathbb{Z}_+^2$.

Let us investigate the limit of $f'_{2,n}$. Next, we note that given (X_k, Y_k) and the event $\mathcal{E}_k = \{12 \in E(G_{n,k})\}$, the random variables $D_{1,k}$ and $D_{2,k}$ are independent, and both distributed according to $1 + \text{Bin}(X_k - 2, Y_k)$. Hence

$$
\begin{aligned}
&\mathbb{P}(D_{1,k} = s, D_{2,k} = t, \mathcal{E}_k \mid X_k, Y_k) \\
&= \mathbb{P}(\mathcal{E}_k \mid X_k, Y_k)\, \mathbb{P}(D_{1,k} = s, D_{2,k} = t \mid \mathcal{E}_k, X_k, Y_k) \\
&= \frac{(X_k)_2}{(n)_2} Y_k\, \text{Bin}(X_k - 2, Y_k)(s - 1)\, \text{Bin}(X_k - 2, Y_k)(t - 1).
\end{aligned}
$$

By taking expectations above, and dividing the outcome by $\mathbb{P}(\mathcal{E}_k) = \mathbb{E} \frac{(X_k)_2}{(n)_2} Y_k = (n)_2^{-1} P_{21}^{(n)}$, it follows that

$$f'_{2,n}(s,t) = \int \mathrm{Bin}(x-2,y)(s-1)\,\mathrm{Bin}(x-2,y)(t-1) \frac{(x)_2 y\, P^{(n)}(dx,dy)}{P_{21}^{(n)}}.$$

When $P^{(n)} \to P$ weakly and $P_{21}^{(n)} \to P_{21} \in (0,\infty)$, it follows that $f'_{2,n}(s,t) \to f'_2(s-1,t-1)$ pointwise on $\mathbb{Z}_+^2$, where f'_2 is defined by (8). Hence

$$(f_n \otimes f_n) * f'_{2,n} \to \delta_{(1,1)} * (f \otimes f) * f'_2$$

pointwise, and together with (13), we conclude that Theorem 1:(i) is valid. $\qquad\square$

6.3   Proof of Theorem 1:(ii)

The proof is similar to the proof of [39, Theorem 3.2], but slightly simpler because here we analyze model distributions instead of empirical distributions of random graph samples. Let $(D_{1,n}^*, D_{2,n}^*) \in \mathbb{Z}_+^2$ be a random variable distributed according to the model bidegree distribution $f_{2,n}$ of $G^{(n)}$. Theorem 1:(i) states that $(D_{1,n}^*, D_{2,n}^*) \to (D_1^*, D_2^*)$ weakly. Now let $\phi : \mathbb{Z}_+^2 \to \mathbb{R}$ be a function bounded by $|\phi(x,y)| \le c(1+x^2+y^2)$. Skorohod's coupling theorem [22, Theorem 4.30] implies that there exist a probability space and some random variables $(\tilde{D}_{1,n}^*, \tilde{D}_{2,n}^*) =_{\mathrm{st}} (D_{1,n}^*, D_{2,n}^*)$ and $(\tilde{D}_1^*, \tilde{D}_2^*) =_{\mathrm{st}} (D_1^*, D_2^*)$ such that $(\tilde{D}_{1,n}^*, \tilde{D}_{2,n}^*) \to (\tilde{D}_1^*, \tilde{D}_2^*)$ almost surely. Then $Z_n := \phi(\tilde{D}_{1,n}^*, \tilde{D}_{2,n}^*) \to \phi(\tilde{D}_1^*, \tilde{D}_2^*) =: Z$ almost surely. Also $|Z_n| \le c(1 + (\tilde{D}_{1,n}^*)^2 + (\tilde{D}_{2,n}^*)^2) =: Z'_n$ a.s. With the help of Lemma 1, we note that

$$\mathbb{E}((\tilde{D}_{1,n}^*)^2) = \frac{\mathbb{E} D_{1,n}^3}{\mathbb{E} D_{1,n}} \to \frac{\mathbb{E} D_1^3}{\mathbb{E} D_1} = \mathbb{E}((D_1^*)^2) < \infty,$$

and hence $\mathbb{E} Z'_n \to \mathbb{E} Z' = c(1 + 2\mathbb{E}((D_1^*)^2)) < \infty$. Lebesgue's dominated convergence theorem (see the version in [22, Theorem 1.21]) now implies that $\mathbb{E} Z_n \to \mathbb{E} Z$, which confirms the claim. $\qquad\square$

6.4   Proof of Theorem 2

We only sketch the proof in the case where $\frac{m}{n} \to \mu \in (0,\infty)$. By applying Theorem 1:(ii) with $\phi(x,y) = x$, and then with $\phi(x,y) = x^2$, we find that $\mathrm{Var}(D_{1,n}^*) \to \mathrm{Var}(D_1^*)$. Observe next that for $\phi(x,y) = xy$, $|\phi(x,y)| \le 2(x^2+y^2)$. Hence Theorem 1:(ii) also implies that $\mathrm{Cov}(D_{1,n}^*, D_{2,n}^*) \to \mathrm{Cov}(D_1^*, D_2^*)$. Hence the claim follows by Proposition 1. $\qquad\square$

6.5   Proof of Theorem 3

Because $f_{2,n}$ has identical marginals, we see that

$$\rho_{\text{Ken}}(f_{2,n}) \;=\; \frac{\int \phi\, d(f_{2,n} \otimes f_{2,n}) - \left(\int \phi_1\, d(f_{2,n}^{(1)} \otimes f_{2,n}^{(1)})\right)^2}{\int \phi_1^2\, d(f_{2,n}^{(1)} \otimes f_{2,n}^{(1)}) - \left(\int \phi_1\, d(f_{2,n}^{(1)} \otimes f_{2,n}^{(1)})\right)^2},$$

where $\phi_1(x_1, y_1) = \text{sgn}(x_1 - y_1)$, $\phi(x_1, x_2, y_1, y_2) = \phi_1(x_1, y_1)\phi_1(x_2, y_2)$ are bounded (and trivially continuous) functions defined on $\mathbb{Z}_+^2$ and $\mathbb{Z}_+^4$, respectively. Theorem 1 implies that $f_{2,n} \to f_2$ weakly as probability measures on $\mathbb{Z}_+^2$. Hence also $f_{2,n} \otimes f_{2,n} \to f_2 \otimes f_2$ and $f_{2,n}^{(1)} \otimes f_{2,n}^{(1)} \to f_2^{(1)} \otimes f_2^{(1)}$ weakly. Hence we conclude that $\rho_{\text{Ken}}(f_{2,n}) \to \rho_{\text{Ken}}(f_2)$.

To verify the claim for Spearman's rank correlation, we apply the representation [33, Section 4.3]

$$\rho_{\text{Spe}}(f_{2,n}) \;=\; \frac{3(\mathbb{P}((X_1 - Y_2)(X_2 - Z_2) > 0) - \mathbb{P}((X_1 - Y_2)(X_2 - Z_2) < 0))}{\sqrt{1 - \mathbb{P}(X_1 = Y_1 = Z_1)}\sqrt{1 - \mathbb{P}(X_2 = Y_2 = Z_2)}},$$

where $(X_1, X_2), (Y_1, Y_2), (Z_1, Z_2)$ are mutually independent and $f_{2,n}$-distributed. Because $f_{2,n}$ has identical marginals, this can be rewritten as

$$\rho_{\text{Spe}}(f_{2,n}) \;=\; 3\frac{\int \phi\, d(f_{2,n} \otimes f_{2,n} \otimes f_{2,n})}{\int \psi\, d(f_{2,n}^{(1)} \otimes f_{2,n}^{(1)} \otimes f_{2,n}^{(1)})},$$

where $\phi(x_1, x_2, y_1, y_2, z_1, z_2) = \text{sgn}((x_1 - y_2)(x_2 - z_2))$ and $\psi(x_1, y_1, z_1) = 1 - 1(x_1 = y_1 = z_1)$ are bounded (and trivially continuous) functions on $\mathbb{Z}_+^6$ and $\mathbb{Z}_+^3$, respectively. The second claim follows by noting that $f_{2,n} \otimes f_{2,n} \otimes f_{2,n} \to f_2 \otimes f_2 \otimes f_2$ and $f_{2,n}^{(1)} \otimes f_{2,n}^{(1)} \otimes f_{2,n}^{(1)} \to f_2^{(1)} \otimes f_2^{(1)} \otimes f_2^{(1)}$ weakly. □

Lemma 1. *Assume that $P^{(n)} \to P_{rs}$ weakly and $P_{rs}^{(n)} \to P_{rs} < \infty$ for $rs = 10, 21, 32, 43$, with $P_{10} > 0$. Then the third moments of the model degree distribution converge according to $\sum_s s^3 f_n(s) \to \sum_s s^3 f(s) < \infty$.*

Proof. To appear in extended version.

References

1. Abbe, E.: Community detection and stochastic block models: recent developments. J. Mach. Learn. Res. **18**, 1–86 (2018)
2. Aiello, W., Bonato, A., Cooper, C., Janssen, J., Prałat, P.: A spatial web graph model with local influence regions. Internet Math. **5**(1–2), 175–196 (2008)
3. Albert, R., Barabási, A.L.: Statistical mechanics of complex networks. Rev. Mod. Phys. **74**, 47–97 (2002). https://doi.org/10.1103/RevModPhys.74.47
4. Amerise, I.L., Tarsitano, A.: Correction methods for ties in rank correlations. J. Appl. Stat. **42**(12), 2584–2596 (2015). https://doi.org/10.1080/02664763.2015. 1043870

5. Ángeles Serrano, M., Boguñá, M.: Clustering in complex networks I. General formalism. Phys. Rev. E **74**, 056114 (2006). https://doi.org/10.1103/PhysRevE.74.056114
6. Ball, F.G., Sirl, D.J., Trapman, P.: Epidemics on random intersection graphs. Ann. Appl. Probab. **24**(3), 1081–1128 (2014). https://doi.org/10.1214/13-AAP942
7. Bloznelis, M.: Degree and clustering coefficient in sparse random intersection graphs. Ann. Appl. Probab. **23**(3), 1254–1289 (2013). https://doi.org/10.1214/12-AAP874
8. Bloznelis, M.: Degree-degree distribution in a power law random intersection graph with clustering. Internet Math. 1–25 (2017)
9. Bloznelis, M., Jaworski, J., Kurauskas, V.: Assortativity and clustering of sparse random intersection graphs. Electron. J. Probab. **18**(38), 24 (2013). https://doi.org/10.1214/EJP.v18-2277
10. Bloznelis, M., Leskelä, L.: Clustering and percolation on superpositions of Bernoulli random graphs (2019). https://arxiv.org/abs/1912.13404
11. Bode, M., Fountoulakis, N., Müller, T.: On the largest component of a hyperbolic model of complex networks. Electron. J. Comb. **22**(3), 1–52 (2015)
12. Boguñá, M., Pastor-Satorras, R.: Class of correlated random networks with hidden variables. Phys. Rev. E **68**, 036112 (2003). https://doi.org/10.1103/PhysRevE.68.036112
13. Bollobás, B., Janson, S., Riordan, O.: The phase transition in inhomogeneous random graphs. Random Struct. Algor. **31**(1), 3–122 (2007). https://doi.org/10.1002/rsa.20168
14. Breiger, R.L.: The duality of persons and groups. Soc. Forces **53**(2), 181–190 (1974). https://doi.org/10.1093/sf/53.2.181
15. Britton, T., Deijfen, M., Lagerås, A.N., Lindholm, M.: Epidemics on random graphs with tunable clustering. J. Appl. Probab. **45**(3), 743–756 (2008). https://doi.org/10.1239/jap/1222441827
16. Czabarka, É., Rauh, J., Sadeghi, K., Short, T., Székely, L.: On the number of non-zero elements of joint degree vectors. Electron. J. Comb. **24**(1), 1–14 (2017)
17. Frieze, A., Karoński, M.: Introduction to Random Graphs. Cambridge University Press, Cambridge (2015). https://doi.org/10.1017/CBO9781316339831
18. Godehardt, E., Jaworski, J.: Two models of random intersection graphs and their applications. Electron. Notes Discret. Math. **10**, 129–132 (2001)
19. Gröhn, T., Karjalainen, J., Leskelä, L.: Clique and cycle frequencies in a sparse random graph model with overlapping communities, November 2019. https://arxiv.org/abs/1911.12827
20. Holland, P.W., Laskey, K.B., Leinhardt, S.: Stochastic blockmodels: first steps. Soc. Netw. **5**(2), 109–137 (1983). https://doi.org/10.1016/0378-8733(83)90021-7
21. Jacob, E., Mörters, P.: Robustness of scale-free spatial networks. Ann. Probab. **45**(3), 1680–1722 (2017). https://doi.org/10.1214/16-AOP1098
22. Kallenberg, O.: Foundations of Modern Probability. APA. Springer, New York (2002). https://doi.org/10.1007/978-1-4757-4015-8
23. Karjalainen, J., van Leeuwaarden, J.S.H., Leskelä, L.: Parameter estimators of sparse random intersection graphs with thinned communities. In: Bonato, A., Prałat, P., Raigorodskii, A. (eds.) WAW 2018. LNCS, vol. 10836, pp. 44–58. Springer, Cham (2018). https://doi.org/10.1007/978-3-319-92871-5_4
24. Karjalainen, J., Leskelä, L.: Moment-based parameter estimation in binomial random intersection graph models. In: Bonato, A., Chung Graham, F., Prałat, P. (eds.) WAW 2017. LNCS, vol. 10519, pp. 1–15. Springer, Cham (2017). https://doi.org/10.1007/978-3-319-67810-8_1

25. Karoński, M., Scheinerman, E.R., Singer-Cohen, K.B.: On random intersection graphs: the subgraph problem. Comb. Probab. Comput. **8**(1–2), 131–159 (1999). https://doi.org/10.1017/S0963548398003459
26. Kiwi, M., Mitsche, D.: On the second largest component of random hyperbolic graphs. SIAM J. Discret. Math. **33**(4), 2200–2217 (2019). https://doi.org/10.1137/18M121201X
27. Krioukov, D., Papadopoulos, F., Kitsak, M., Vahdat, A., Boguñá, M.: Hyperbolic geometry of complex networks. Phys. Rev. E **82**(3), 036106 (2010)
28. Krot, A., Prokhorenkova, L.O.: Assortativity in generalized preferential attachment models. Internet Math. 9–21 (2017)
29. Kruskal, W.H.: Ordinal measures of association. J. Am. Stat. Assoc. **53**(284), 814–861 (1958). http://www.jstor.org/stable/2281954
30. Kurauskas, V.: On local weak limit and subgraph counts for sparse random graphs (2015). https://arxiv.org/abs/1504.08103
31. Mahadevan, P., Krioukov, D., Fall, K., Vahdat, A.: Systematic topology analysis and generation using degree correlations. In: Proceedings of the 2006 Conference on Applications, Technologies, Architectures, and Protocols for Computer Communications, SIGCOMM 2006, pp. 135–146 (2006). https://doi.org/10.1145/1159913.1159930
32. Molloy, M., Reed, B.: The size of the giant component of a random graph with a given degree sequence. Comb. Probab. Comput. **7**(3), 295–305 (1998). https://doi.org/10.1017/S0963548398003526
33. Nešlehová, J.: On rank correlation measures for non-continuous random variables. J. Multivar. Anal. **98**(3), 544–567 (2007). https://doi.org/10.1016/j.jmva.2005.11.007
34. Newman, M.E.J.: Assortative mixing in networks. Phys. Rev. Lett. **89**, 208701 (2002). https://doi.org/10.1103/PhysRevLett.89.208701
35. Newman, M.E.J.: The structure and function of complex networks. SIAM Rev. **45**(2), 167–256 (2003). https://doi.org/10.1137/S003614450342480
36. Petti, S., Vempala, S.: Approximating sparse graphs: the random overlapping communities model (2018). https://arxiv.org/abs/1802.03652
37. Sadeghi, K., Rinaldo, A.: Statistical models for degree distributions of networks. In: NIPS (2014)
38. Vadon, V., Komjáthy, J., van der Hofstad, R.: A new model for overlapping communities with arbitrary internal structure. Appl. Netw. Sci. **4**(1), 1–19 (2019). https://doi.org/10.1007/s41109-019-0149-9
39. van der Hofstad, R., Litvak, N.: Degree-degree dependencies in random graphs with heavy-tailed degrees. Internet Math. **10**(3), 287–334 (2014). https://doi.org/10.1080/15427951.2013.850455
40. van der Hoorn, P., Litvak, N.: Convergence of rank based degree-degree correlations in random directed networks. Moscow J. Comb. Num. Theory **4**(4), 427–465 (2014)
41. van der Hoorn, P., Litvak, N.: Degree-degree dependencies in directed networks with heavy-tailed degrees. Internet Math. **11**(2), 155–179 (2015)
42. Vázquez, A., Pastor-Satorras, R., Vespignani, A.: Large-scale topological and dynamical properties of the internet. Phys. Rev. E **65**, 066130 (2002). https://doi.org/10.1103/PhysRevE.65.066130
43. Villani, C.: Optimal Transport: Old and New. Springer, New York (2009)
44. Yang, J., Leskovec, J.: Structure and overlaps of ground-truth communities in networks. ACM Trans. Intell. Syst. Technol. **5**(2) (2014). https://doi.org/10.1145/2594454

Clustering Coefficient of a Preferred Attachment Affiliation Network

Daumilas Ardickas and Mindaugas Bloznelis[✉]

Institute of Informatics, Vilnius University, Naugarduko 24, 03225 Vilnius, Lithuania
mindaugas.bloznelis@mif.vu.lt

Abstract. It is well known that the global clustering coefficient of a standard preferential attachment random graph vanishes as the number of vertices tends to ∞. We evaluate the global clustering coefficient of the preferred attachment affiliation network [4] and show that it is bounded away from zero.

Keywords: Preferential attachment · Affiliation network · Clustering coefficient · Power law · Random intersection graph

1 Introduction and Results

In a preferential attachment network each newly arrived vertex is attached preferentially to already well connected sites, [2]. An important class of social networks are affiliation networks: members of a network tend to establish relations if they share some common features [8,14]. For example, customers of a video-sharing website are considered related to each other if they have downloaded the same movie. Here the rich get richer principle [2] affects customers and movies simultaneously: a newly arrived customer u' is likely to download popular items, thus, further increasing their popularity. Similarly, by selecting a highly popular movie, the new customer u' becomes adjacent to highly connected customers (those that have downloaded this movie), thus further, increasing the number of their neighbours.

In the preferred attachment affiliation network [4] vertices (customers) are linked to items (movies) independently at random, and the probability of a link between the new vertex u' and an item w is proportional to the number of vertices already linked to w. Two vertices of the network are declared adjacent whenever there is an item linked to both of them. The network admits (asymptotic) power law degree sequence, see [4].

Model. Let $k > 0$, $l \geq 0$ be integers. Let $0 < \lambda \leq k + l$. Consider an electronic library containing items $w_1, \ldots, w_l$ at the beginning. Every w_j is prescribed initial score $s(w_j) = 1$. On the first step new items $w_{l+1}, \ldots, w_{l+k}$ arrive to the library, each having score 1. Then the first customer v_1 visits the library and downloads items independently at random: An item w is downloaded with probability $p_{1,s(w)} = \lambda s(w)(l+k)^{-1}$. Every item downloaded by v_1 increases its score

© Springer Nature Switzerland AG 2020
B. Kamiński et al. (Eds.): WAW 2020, LNCS 12091, pp. 82–95, 2020.
https://doi.org/10.1007/978-3-030-48478-1_6

by one. Let $W_n = \{w_1, \ldots, w_{l+nk}\}$ be the collection of items in the library after n steps. On the $n+1$ step k new items arrive to the library, each having score 1. Then customer v_{n+1} visits the library and downloads items independently at random: An item w is downloaded with probability

$$p_{n+1,s(w)} = \lambda s(w)(l + (n+1)k + n\lambda)^{-1},$$

which is proportional to the current score $s(w)$ of w. Every item downloaded by v_{n+1} increases its score by one. Note that after n-th step $s(w) - 1$ is the number of customers from $V_n = \{v_1, \ldots, v_n\}$ that have downloaded item w. The preferred attachment affiliation network G_n has vertex set V_n. Two vertices (customers) are adjacent in G_n whenever there is at least one item downloaded by (=linked to) both of them. We note that the expected number of items downloaded by each customer is the same and equals λ.

For convenience, we may represent items as bins. Each newly arrived bin contains a single ball. A new customer v_{n+1} throws balls into bins $w_1, \ldots, w_{l+(n+1)k}$ at random: Each bin w receives a ball with probability $p_{n+1,s(w)}$ and independently of the other bins. The score $s(w)$ counts the (current) number of balls in the bin w. This number may increase with n. It measures the popularity of the bin (item) w. Hence, popular bins (items) have higher chances to be selected.

Clustering Coefficient. The global clustering coefficient of a non-random finite graph $\mathcal{G}$ is the conditional probability $C(\mathcal{G}) = \mathbf{P}\{v_2^* \sim v_3^* | v_1^* \sim v_2^*, v_1^* \sim v_3^*\}$. Here (v_1^*, v_2^*, v_3^*) is an (ordered) vertex triple sampled uniformly at random and $\sim$ stands for the adjacency relation. The calculation of $C(\mathcal{G})$ reduces to the subgraph counts $N_\triangle$ (= the number of triangles in $\mathcal{G}$) and $N_\vee$ (= the number of paths of length 2 in $\mathcal{G}$). Indeed, for a non-random graph $\mathcal{G}$ we have $C(\mathcal{G}) = 3N_\triangle/N_\vee$.

In the case of random graph G_n the conditional probability

$$C(G_n) = \mathbf{P}\{v_2^* \sim v_3^* | v_1^* \sim v_2^*, v_1^* \sim v_3^*\} \tag{1}$$

refers to the two sources of randomness: the random graph generation and the sampling of the vertex triple. We assume that the vertex triple (v_1^*, v_2^*, v_3^*) (sampled uniformly at random from V_n) is independent of G_n.

Results. In Theorem 1 we establish a first order approximation to the clustering coefficient $C(G_n)$ as $n \to +\infty$. Let

$$\alpha = \frac{k}{\lambda}, \quad \beta = \frac{1}{1+\alpha}, \quad \gamma = \frac{k\beta(1-3\beta)}{9(1-2\beta)}\left(\frac{3}{1-\beta} + \frac{2}{1-2\beta}\right), \quad \tau = \frac{2k\beta(1-3\beta)}{3(1-2\beta)^2}.$$

Theorem 1. *Let $n \to +\infty$. We have*

$$C(G_n) = \begin{cases} 1 - o(1), & \text{for } \alpha \le 2, \\ (1+\gamma)^{-1} + o(1), & \text{for } \alpha > 2. \end{cases} \tag{2}$$

In the proof of Theorem 1 we establish a first order asymptotics to the expected number of triangles and paths of length 2 ("cherries"), see (44) and (46), (47) below.

The random graph G_n is defined by the bipartite graph with bipartition $V_n \cup W_n$, where customers (vertices) $v \in V_n$ are linked to movies (items) $w \in W_n$ they have downloaded. G_n is related to the passive random intersection graph introduced in [6], where items receive weights (=scores) independently at random and, given the weights, vertices are linked to items independently at random with probabilities proportional to the weights. Two vertices of the intersection graph are declared adjacent whenever there is an item linked to both of them. Depending on the distribution of random weights, the global clustering coefficient of passive random intersection graph is $1 - o(1)$ when the weights have infinite third moment, and it is less than one when weights have a finite third moment, [3,7]. Theorem 1 reveals a similar pattern. Indeed, it has been shown in [4] that (in the preferred attachment affiliation network model) the fraction of items with score i scales as $i^{-2-\alpha}$. Therefore the (asymptotic) score sequence has infinite third moment whenever $\alpha \leq 2$.

Vertices $v_1, \ldots, v_n$ of G_n are numbered in the order of their arrival. It is interesting to know whether and how the arrival time of a vertex affects the local clustering characteristics in a vicinity of that vertex. For this purpose we study the conditional probabilities

$$C_{x,y,z} = \mathbf{P}\{v_x \sim v_z | v_x \sim v_y \sim v_z\}, \qquad x, y, z \in [n], \qquad x \neq y \neq z.$$

For $s < t < u$, let

$$\gamma_{t,s,u} = \tau, \qquad \gamma_{s,t,u} = \tau s^\beta t^{-\beta}, \qquad \gamma_{s,u,t} = \tau s^\beta t^{1-2\beta} u^{\beta-1},$$

In Theorem 2 below we assume that $s = s_n$, $t = t_n$ and $u = u_n$.

Theorem 2. *Let $n \to +\infty$. For $1 < s < t < u$ such that $s \to +\infty$ we have for each triple $(x, y, z) \in \{(s, t, u), (t, s, u), (s, u, t)\}$ that*

$$C_{x,y,z} = \begin{cases} 1 - o(1), & \text{for } \alpha \leq 2, \\ (1 + \gamma_{x,y,z})^{-1} + o(1), & \text{for } \alpha > 2. \end{cases} \tag{3}$$

The results of Theorem 2 might be applicable to link prediction problems and recommender systems in dynamic settings.

Related Work. It has already been mentioned that in a standard preferential attachment (PA) random graph the global clustering coefficient vanishes. The clustering property of the PA affiliation network considered in the present paper is facilitated by the underlying bipartite structure (vertices from V_n are linked to items from W_n). Similarly, the clustering property has been established in the spatial PA model [1] and random intersection graph process [5], where the underlying bipartite structure relates vertices to regions (subsets) of a given domain (set), see [5,10,11]. Generally, the relation between the bipartite structure and clustering property in social networks has been discussed in [14]. Clustering properties of a real co-authorship (affiliation) network evolving in time have been reported in [13]. We also mention an approach of [9,15], where the clustering property of PA type random graph is enhanced by inserting (at each step) extra edges that create triangles. Another model of evolving affiliation network has been considered in [12], but the clustering property has not been addressed.

2 Proofs

The section is organized as follows. Before the proofs we introduce notation. Then in two technical lemmas we present auxiliary results. Afterwards we give proofs of Theorems 1 and 2.

Notation. Given $w \in W_{i+1}$, let $S_i(w)$ be the score of w just before the arrival of the user v_{i+1}. In particular the score of $w \in W_{i+1} \setminus W_i$ is $S_i(w) = 1$. For $m = 1, 2, 3$, let

$$\mu_t^{(m)} = \mathbb{E}\left(\sum_{w \in W_t} S_{t-1}^m(w) \right).$$

Given vertex v_u and item w we denote by $A_{v_u, w} = \{v_u \to w\}$ the event that user v_u has downloaded item w. We write, for short,

$$A_{st,w} = A_{v_s,w} \cap A_{v_t,w}, \qquad A_{stu,w} = A_{v_s,w} \cap A_{v_t,w} \cap A_{v_u,w}.$$

The event $A_{st,w}$ implies that vertices v_s and v_t are adjacent. We call w a witness of the edge $v_s \sim v_t$ whenever $A_{st,w}$ occurs. For $1 \le s < t$ the union $\cup_{w \in W_s} A_{st,w} = \{v_s \sim v_t\}$ is the event that v_s and v_t are adjacent. For $1 \le s < t < u$ let $\Delta_{stu} = \{v_s \sim v_t, v_s \sim v_u, v_t \sim v_u\}$ denote the event that vertices v_s, v_t, v_u make up a triangle in G_n. Furthermore, let

$$\Delta_{stu}^* = \bigcup_{w \in W_s} A_{stu,w}$$

be the event that some item is shared by the three vertices v_s, v_t, v_u. Clearly, the event Δ_{stu}^* implies event Δ_{stu}. Let $\vee_{stu} = \{v_s \sim v_t, v_t \sim v_u\}$ denote the event that v_s and v_u are neighbours of v_t.

By $\theta, \theta', \theta''$ we denote numbers with absolute value bounded by a constant that only depends on k, l and λ. The numbers $\theta, \theta', \theta''$ may take different values in different places.

2.1   Auxiliary Results

Lemma 1. *Let $n, t \to +\infty$. Assume that $1 < t \le n$. Then there exist constants $c_1, c_2 > 0$ depending on k, l, λ such that*

$$\mu_t^{(1)} = (k + \lambda)t + O(t^\beta) = \frac{k}{1 - \beta}t + O(t^\beta), \tag{4}$$

$$\mu_t^{(2)} = \begin{cases} k\left(\frac{2}{1-2\beta} - \frac{1}{1-\beta}\right)t + O(t^{2\beta}), & \text{for} \quad 2\beta < 1, \\ 2kt\ln t + O(t), & \text{for} \quad 2\beta = 1, \\ (c_1 + o(1))t^{2\beta}, & \text{for} \quad 2\beta > 1, \end{cases} \tag{5}$$

$$\mu_t^{(3)} = \begin{cases} k\left(\frac{6}{1-3\beta} - \frac{6}{1-2\beta} + \frac{1}{1-\beta}\right)t + O(t^{3\beta}), & \text{for} \quad 3\beta < 1, \\ 6kt\ln t + O(t), & \text{for} \quad 3\beta = 1, \\ (c_2 + o(1))t^{3\beta}, & \text{for} \quad 3\beta > 1. \end{cases} \tag{6}$$

Proof of Lemma 1. Let $l^* = \alpha + l/\lambda$. We have

$$p_{i+1,x} = \frac{x}{i(\alpha + 1) + l^*} = x\left(\frac{\beta}{i} - R_i\right), \qquad 0 < R_i < (\beta/i)^2 l^*. \tag{7}$$

For $k = 1, 2, 3$ and $i \le j - 1$ we put

$$u_{i,j}^{(k)} = \prod_{s=i}^{j-1}\left(1 + \frac{k\beta}{s} - kR_s\right)$$

and define $u_{i,i}^{(k)} = 1$. We have for $1 \le i \le j$

$$u_{i,j}^{(k)} = \frac{j^{k\beta}}{i^{k\beta}}\left(1 + \frac{\theta}{i}\right). \tag{8}$$

Proof of (4). For $w \in W_{i+1}$, $i \ge 1$, relation (7) implies

$$\mathbb{E}(S_{i+1}(w)|S_i(w) = x) = x + p_{i+1,x} = x\left(1 + \frac{\beta}{i} - R_i\right). \tag{9}$$

By iterating (9), we obtain for $w \in W_{i+1}$ and $1 \le i < j$

$$\mathbb{E}(S_j(w)|S_i(w) = x) = u_{i,j}^{(1)}x = \frac{j^\beta}{i^\beta}\left(1 + \frac{\theta}{i}\right)x. \tag{10}$$

In the case where $w \in W_{i+1} \setminus W_i$ we have $S_i(w) \equiv 1$ and therefore

$$\mathbb{E}S_j(w) = \frac{j^\beta}{i^\beta}\left(1 + \frac{\theta}{i}\right). \tag{11}$$

Finally,

$$\mu_t^{(1)} = \sum_{w \in W_1} \mathbb{E} S_{t-1}(w) + \sum_{i=1}^{t-1} \sum_{w \in W_{i+1} \setminus W_i} \mathbb{E} S_{t-1}(w) \tag{12}$$

$$= (k+l)t^\beta(1+\theta) + k \sum_{i=1}^{t-1} \frac{(t-1)^\beta}{i^\beta} \left(1 + \frac{\theta}{i}\right)$$

$$= \frac{k}{1-\beta} t + O(t^\beta).$$

Proof of (5). For $w \in W_{i+1}$, $i \geq 1$, relation (7) implies

$$\mathbb{E}(S_{i+1}^2(w)|S_i(w) = x) = x^2(1 - p_{i+1,x}) + (x+1)^2 p_{i+1,x} \tag{13}$$

$$= x^2 \left(1 + \frac{2\beta}{i} - 2R_i\right) + x \left(\frac{\beta}{i} - R_i\right).$$

By iterating (13), we obtain for $w \in W_{i+1}$ and $1 \leq i < j$

$$\mathbb{E}(S_j^2(w)|S_i(w) = x) = u_{i,j}^{(2)} x^2 + \sum_{t=i}^{j-1} \left(\frac{\beta}{t} - R_t\right) u_{t+1,j}^{(2)} \mathbb{E}(S_t(w)|S_i(w) = x). \tag{14}$$

Now (8) and (10) imply

$$\mathbb{E}(S_j^2(w)|S_i(w) = x) = \frac{j^{2\beta}}{i^{2\beta}} \left(1 + \frac{\theta}{i}\right) x^2 + h_{i,j} x, \tag{15}$$

$$h_{i,j} = \sum_{t=i}^{j-1} \left(\frac{\beta}{t} - R_t\right) \frac{j^{2\beta}}{t^{2\beta}} \left(1 + \frac{\theta}{t}\right) \frac{t^\beta}{i^\beta} \left(1 + \frac{\theta'}{i}\right).$$

A straightforward calculation shows

$$h_{i,j} = \frac{j^{2\beta}}{i^{2\beta}} \left(1 + \frac{\theta}{i}\right) - \frac{j^\beta}{i^\beta} \left(1 + \frac{\theta'}{i}\right). \tag{16}$$

Hence,

$$\mathbb{E} S_j^2(w) = \frac{j^{2\beta}}{i^{2\beta}} \left(1 + \frac{\theta}{i}\right) \mathbb{E} S_i^2(w) \tag{17}$$

$$+ \left(\frac{j^{2\beta}}{i^{2\beta}} \left(1 + \frac{\theta'}{i}\right) - \frac{j^\beta}{i^\beta} \left(1 + \frac{\theta''}{i}\right)\right) \mathbb{E} S_i(w).$$

For $w \in W_{i+1} \setminus W_i$ we have $S_i(w) \equiv 1$. Now (17) implies

$$\mathbb{E} S_j^2(w) = 2 \frac{j^{2\beta}}{i^{2\beta}} \left(1 + \frac{\theta}{i}\right) - \frac{j^\beta}{i^\beta} \left(1 + \frac{\theta'}{i}\right). \tag{18}$$

From this identity we derive (5) proceeding similarly as in (12) above.

Proof of (6). For $w \in W_{i+1}$, $i \geq 1$, relation (7) implies

$$\mathbb{E}(S_{i+1}^3(w)|S_i(w) = x) = x^3(1 - p_{i+1,x}) + (x+1)^3 p_{i+1,x} \tag{19}$$

$$= x^3\left(1 + \frac{3\beta}{i} - 3R_i\right) + (3x^2 + x)\left(\frac{\beta}{i} - R_i\right).$$

By iterating (19), we obtain for $w \in W_{i+1}$ and $1 \leq i < j$

$$\mathbb{E}(S_j^3(w)|S_i(w) = x) = u_{i,j}^{(3)} x^3$$

$$+ \sum_{t=i}^{j-1}\left(\frac{\beta}{t} - R_t\right) u_{t+1,j}^{(3)}\mathbb{E}(3S_t^2(w) + S_t(w)|S_i(w) = x).$$

Now (8), (10) and (15), (16) imply

$$\mathbb{E}(S_j^3(w)|S_i(w) = x) = \frac{j^{3\beta}}{i^{3\beta}}\left(1 + \frac{\theta}{i}\right)x^3 + 3h_{i,j}^{(1)}x^2 + h_{i,j}^{(2)}x, \tag{20}$$

where

$$h_{i,j}^{(1)} = \sum_{t=i}^{j-1}\left(\frac{\beta}{t} - R_t\right)u_{t+1,j}^{(3)}\frac{t^{2\beta}}{i^{2\beta}}\left(1 + \frac{\theta}{i}\right) = \frac{j^{3\beta}}{i^{3\beta}}\left(1 + \frac{\theta}{i}\right) - \frac{j^{2\beta}}{i^{2\beta}}\left(1 + \frac{\theta'}{i}\right),$$

$$h_{i,j}^{(2)} = \sum_{t=i}^{j-1}\left(\frac{\beta}{t} - R_t\right)u_{t+1,j}^{(3)}\left(3\frac{t^{2\beta}}{i^{2\beta}}\left(1 + \frac{\theta'}{i}\right) - 2\frac{t^{\beta}}{i^{\beta}}\left(1 + \frac{\theta''}{i}\right)\right)$$

$$= 2\frac{j^{3\beta}}{i^{3\beta}}\left(1 + \frac{\theta}{i}\right) - 3\frac{j^{2\beta}}{i^{2\beta}}\left(1 + \frac{\theta'}{i}\right) + \frac{j^{\beta}}{i^{\beta}}\left(1 + \frac{\theta''}{i}\right).$$

Hence

$$\mathbb{E}(S_j^3(w)|S_i(w)) = \frac{j^{3\beta}}{i^{3\beta}}\left(1 + \frac{\theta}{i}\right)\mathbb{E}S_i^3(w) + 3h_{i,j}^{(1)}\mathbb{E}S_i^2(w) + h_{i,j}^{(2)}\mathbb{E}S_i(w). \tag{21}$$

For $w \in W_{i+1} \setminus W_i$ we have $S_i(w) \equiv 1$. Now (21) implies

$$\mathbb{E}(S_j^3(w)) = 6\frac{j^{3\beta}}{i^{3\beta}}\left(1 + \frac{\theta}{i}\right) - 6\frac{j^{2\beta}}{i^{2\beta}}\left(1 + \frac{\theta'}{i}\right) + \frac{j^{\beta}}{i^{\beta}}\left(1 + \frac{\theta''}{i}\right). \tag{22}$$

This identity yields (6). $\square$

Lemma 2. Let $1 < s < t < u$ be integers. For $s \to +\infty$ we have

$$\mathbf{P}\{v_t \sim v_u\} = (1 + o(1))\beta^2\frac{u^{\beta-1}}{t^{\beta+1}}\left(\mu_t^{(2)} + \mu_t^{(1)}\right), \tag{23}$$

$$\mathbf{P}\{\Delta_{stu}^*\} = (1 + o(1))\beta^3\frac{(ut)^{\beta-1}}{s^{2\beta+1}}\left(\mu_s^{(3)} + 3\mu_s^{(2)} + 2\mu_s^{(1)}\right). \tag{24}$$

Proof of Lemma 2. Proof of (23). For $1 \leq i \leq t < u$ and $w \in W_i \setminus W_{i-1}$ we have, by the total probability formula,

$$\mathbf{P}\{\mathcal{A}_{tu,w}|S_{t-1}(w) = z\} = p_{t,z}\mathbf{P}\{\mathcal{A}_{vu,w}|S_t(w) = z+1\}$$

$$= p_{t,z} \sum_{y \geq z+1} \mathbf{P}\{\mathcal{A}_{vu,w}|S_{u-1}(w) = y\}\mathbf{P}\{S_{u-1} = y|S_t(w) = z+1\}$$

$$= p_{t,z} \sum_{y \geq z+1} p_{u,y}\mathbf{P}\{S_{u-1}(w) = y|S_t(w) = z+1\}$$

$$= p_{t,z} \frac{\lambda}{l + uk + (u-1)\lambda} \sum_{y \geq z+1} y\mathbf{P}\{S_{u-1} = y|S_t(w) = z+1\}$$

$$= p_{t,z} \frac{\lambda}{l + uk + (u-1)\lambda} \mathbb{E}(S_{u-1}(w)|S_t(w) = z+1)$$

$$= \beta^2 \frac{z(z+1)}{ut} \frac{u^\beta}{t^\beta} \left(1 + \frac{\theta}{t}\right). \tag{25}$$

In the last step we invoked (10) and used, see (7),

$$p_{t,z} \frac{\lambda}{l + uk + (u-1)\lambda} = z\left(\frac{\beta}{t-1} - \frac{\theta}{t^2}\right)\left(\frac{\beta}{u-1} - \frac{\theta'}{u^2}\right)$$

$$= \beta^2 \frac{z}{ut}\left(1 + \frac{\theta}{t}\right)\left(1 + \frac{\theta'}{u}\right)$$

$$= \beta^2 \frac{z}{ut}\left(1 + \frac{\theta}{t}\right).$$

Using (25) we evaluate the probability

$$\mathbf{P}\{\mathcal{A}_{tu,w}\} = \sum_{z \geq 1} \mathbf{P}\{\mathcal{A}_{tu,w}|S_{t-1}(w) = z\}\mathbf{P}\{S_{t-1}(w) = z\}$$

$$= \beta^2 \frac{u^{\beta-1}}{t^{\beta+1}}\left(1 + \frac{\theta}{t}\right)\mathbb{E}\left(S_{t-1}^2(w) + S_{t-1}(w)\right). \tag{26}$$

Furthermore, using inclusion-exclusion, we obtain as $t \to +\infty$

$$\mathbf{P}\{v_t \sim v_u\} = \mathbf{P}\left\{\bigcup_{w \in W_t} \mathcal{A}_{ut,w}\right\} = (1 + o(1))\,\mathbf{H}_{t,u}, \tag{27}$$

$$\mathbf{H}_{t,u} := \sum_{w \in W_t} \mathbf{P}\{\mathcal{A}_{ut,w}\} = \beta^2 \frac{u^{\beta-1}}{t^{\beta+1}}\left(1 + \frac{\theta}{t}\right)\mathbb{E}\left(\mu_t^{(2)} + \mu_t^{(1)}\right). \tag{28}$$

In the last step we used (26). We arrived to (23).

Proof of (24). Denote $p_{s,t,u|w}(y) = \mathbf{P}\{\mathcal{A}_{stu,w}|S_{s-1}(w) = y\}$. For $1 \leq i \leq s < t < u$ and $w \in W_i \setminus W_{i-1}$ we have, by the total probability formula and (25),

$$p_{s,t,u|w}(y) = p_{s,y}\mathbf{P}\{\mathcal{A}_{ut,w}|S_s(w) = y+1\}$$

$$= p_{s,y} \sum_{z \geq y+1} \mathbf{P}\{\mathcal{A}_{ut,w}|S_{t-1}(w) = z\}\mathbf{P}\{S_{t-1}(w) = z|S_s(w) = y+1\}$$

$$= p_{s,y} \sum_{z \geq y+1} \beta^2 \frac{z(z+1)}{ut} \frac{u^\beta}{t^\beta}\left(1 + \frac{\theta}{t}\right)\mathbf{P}\{S_{t-1}(w) = z|S_s(w) = y+1\}$$

Invoking $p_{s,y} = \beta y s^{-1}(1 + \theta/s)$, see (7), we have

$$p_{s,t,u|w}(y) = \beta^3 y \frac{u^{\beta-1}}{st^{\beta+1}}\left(1 + \frac{\theta}{s}\right) \sum_{z \geq y+1} z(z+1)\mathbf{P}\{S_{t-1}(w) = z|S_s(w) = y+1\}$$

$$= \beta^3 y \frac{u^{\beta-1}}{st^{\beta+1}}\left(1 + \frac{\theta}{s}\right)\mathbb{E}\left(S_{t-1}^2(w) + S_{t-1}(w)|S_s(w) = y+1\right)$$

$$= \beta^3 y \frac{u^{\beta-1}}{st^{\beta+1}}\left(1 + \frac{\theta}{s}\right)\frac{t^{2\beta}}{s^{2\beta}}\left((y+1)^2 + (y+1)\right) \qquad (29)$$

$$= \beta^3 \frac{(tu)^{\beta-1}}{s^{2\beta+1}}\left(1 + \frac{\theta}{s}\right)\left(y^3 + 3y^2 + 2y\right).$$

In (29) we used (10) and (15). We apply the total probability formula once again,

$$\mathbf{P}\{\mathcal{A}_{stu,w}\} = \sum_{y \geq 1} p_{s,t,u|w}(y)\mathbf{P}\{S_{s-1}(w) = y\} \qquad (30)$$

$$= \beta^3 \frac{(tu)^{\beta-1}}{s^{2\beta+1}}\left(1 + \frac{\theta}{s}\right)\left(\mathbb{E}S_{s-1}^3(w) + 3\mathbb{E}S_{s-1}^2(w) + 2\mathbb{E}S_{s-1}(w)\right).$$

Furthermore, using inclusion-exclusion, we obtain as $s, u, t \to +\infty$

$$\mathbf{P}\{\Delta_{stu}^*\} = (1 + o(1)) \sum_{w \in W_s} \mathbf{P}\{\mathcal{A}_{stu,w}\}. \qquad (31)$$

Finally, (30), (31) combined with Lemma 1 imply (24). Proofs of (27), (31) will be presented in an extended version of the paper. □

2.2   Proofs of Theorems 1 and 2

Proof of Theorem 2. The result will follow from Lemmas 1 and 2 by establishing the approximation

$$C_{x,y,z} = \frac{\mathbf{P}\{\Delta_{stu}\}}{\mathbf{P}\{V_{xyz}\}} = (1 + o(1))\frac{\mathbf{P}\{\Delta_{stu}^*\}}{\mathbf{P}\{v_x \sim v_y\}\mathbf{P}\{v_y \sim v_z\} + \mathbf{P}\{\Delta_{stu}^*\}}.$$

We sketch the proof of the respective approximations of the numerator and denominator:

$$\mathbf{P}\{\Delta_{stu}\} = (1 + o(1))\mathbf{P}\{\Delta^*_{stu}\}, \tag{32}$$
$$\mathbf{P}\{\vee_{xyz}\} = (1 + o(1))\big(\mathbf{P}\{v_x \sim v_y\}\mathbf{P}\{v_y \sim v_z\} + \mathbf{P}\{\Delta^*_{stu}\}\big). \tag{33}$$

Proof of (32). The event Δ_{stu} occurs whenever either the edges $v_s \sim v_t$, $v_t \sim v_u$, $v_s \sim v_u$ are witnessed by three distinct items (we denote this event by Δ^0_{stu}), or all three edges are witnessed by the same item (the event denoted Δ^*_{stu}). Therefore, $\mathbf{P}\{\Delta_{stu}\} = \mathbf{P}\{\Delta^*_{stu} \cup \Delta^0_{stu}\}$. To show (32) we prove that

$$\mathbf{P}\{\Delta^0_{stu}\} = o\big(\mathbf{P}\{\Delta^*_{stu}\}\big). \tag{34}$$

With $\bar{w}_1, \bar{w}_2, \bar{w}_3$ denoting three distinct witnesses of the edges $v_s \sim v_t$, $v_t \sim v_u$, $v_s \sim v_u$, we have, by the union bound that $\mathbf{P}\{\Delta^0_{stu}\}$ is at most

$$\sum_{\substack{\bar{w}_1 \in W_s}} \sum_{\substack{\bar{w}_2 \in W_t, \\ \bar{w}_2 \neq \bar{w}_1}} \sum_{\substack{\bar{w}_3 \in W_s, \\ \bar{w}_3 \neq \bar{w}_1, \bar{w}_2}} \mathbf{P}\{\mathcal{A}_{st,\bar{w}_1}\}\mathbf{P}\{\mathcal{A}_{tu,\bar{w}_2}\}\mathbf{P}\{\mathcal{A}_{su,\bar{w}_3}\}$$

$$\leq \left(\sum_{\bar{w}_1 \in W_s} \mathbf{P}\{\mathcal{A}_{st,\bar{w}_1}\}\right)\left(\sum_{\bar{w}_2 \in W_t} \mathbf{P}\{\mathcal{A}_{tu,\bar{w}_2}\}\right)\left(\sum_{\bar{w}_3 \in W_s} \mathbf{P}\{\mathcal{A}_{su,\bar{w}_3}\}\right)$$

$$= (1 + o(1))\beta^6 \frac{u^{2\beta-2}}{t^2 s^{2\beta+2}}\big(\mu_s^{(2)} + \mu_s^{(1)}\big)^2\big(\mu_t^{(2)} + \mu_t^{(1)}\big). \tag{35}$$

In the last step we used (28). We compare quantity (35) with expression (24). At this step we use Lemma 1. Now a straightforward calculation shows (34).

Proof of (33). We only consider the case where $x < y < z$, i.e, $x = s, y = t, z = u$. Remaining cases are treated similarly. The event $\{v_s \sim v_t, v_t \sim v_u\}$ occurs whenever the edges $v_s \sim v_t$, $v_t \sim v_u$ are witnessed by two distinct items (we denote this event by $\vee^0_{stu}$), or both edges are witnessed by the same item (this is the event Δ^*_{stu}). Therefore $\mathbf{P}\{v_s \sim v_t, v_t \sim v_u\} = \mathbf{P}\{\Delta^*_{stu} \cup \vee^0_{stu}\}$. By inclusion - exclusion,

$$\mathbf{P}\{\vee_{stu}\} = \mathbf{P}\{\vee^0_{stu}\} + \mathbf{P}\{\Delta^*_{stu}\} - \mathbf{P}\{\vee^0_{stu} \cap \Delta^*_{stu}\}. \tag{36}$$

In the next step we consider the cases $2\beta > 1$ and $2\beta \leq 1$ separately. For $2\beta \leq 1$ we derive (33) from (36) combined with the relations

$$\mathbf{P}\{\vee^0_{stu} \cap \Delta^*_{stu}\} = o\big(\mathbf{P}\{\Delta^*_{stu}\}\big), \tag{37}$$
$$\mathbf{P}\{\vee^0_{stu}\} = (1 + o(1))\mathbf{P}\{v_s \sim v_t\}\mathbf{P}\{v_t \sim v_u\}. \tag{38}$$

Let us show (38). With $\bar{w}_1, \bar{w}_2$ being distinct witnesses of the edges $v_s \sim v_t$, $v_t \sim v_u$, we have

$$\mathbf{P}\{\vee^0_{stu}\} = \mathbf{P}\left\{\bigcup_{\substack{\bar{w}_1 \in W_s, \bar{w}_2 \in W_t \\ \bar{w}_1 \neq \bar{w}_2}} \big(\mathcal{A}_{st,\bar{w}_1} \cap \mathcal{A}_{tu,\bar{w}_2}\big)\right\} \tag{39}$$

$$= (1 + o(1)) \sum_{\bar{w}_1 \in W_s} \sum_{\bar{w}_2 \in W_t \setminus \{\bar{w}_1\}} \mathbf{P}\{\mathcal{A}_{st,\bar{w}_1} \cap \mathcal{A}_{tu,\bar{w}_2}\} \tag{40}$$

and

$$\sum_{\bar{w}_1 \in W_s} \sum_{\bar{w}_2 \in W_t \setminus \{\bar{w}_1\}} \mathbf{P}\{\mathcal{A}_{st,\bar{w}_1} \cap \mathcal{A}_{tu,\bar{w}_2}\}$$

$$= \sum_{\bar{w}_1 \in W_s} \sum_{\bar{w}_2 \in W_t \setminus \{\bar{w}_1\}} \mathbf{P}\{\mathcal{A}_{st,\bar{w}_1}\} \mathbf{P}\{\mathcal{A}_{tu,\bar{w}_2}\} \tag{41}$$

$$= (1 + o(1)) (\mathbf{H}_{s,t} \mathbf{H}_{t,u} - \mathbf{R}).$$

Here $\mathbf{H}_{s,t}$, $\mathbf{H}_{t,u}$ are the same as in (28) and

$$\mathbf{R} = \sum_{w \in W_s} \mathbf{P}\{\mathcal{A}_{st,w}\} \mathbf{P}\{\mathcal{A}_{tu,w}\}.$$

In (40) and (41) we use inclusion-exclusion and the independence of $\mathcal{A}_{st,\bar{w}_1}$ and $\mathcal{A}_{tu,\bar{w}_2}$. Next we upperbound $\mathbf{R}$. From (26) and (11), (18) we obtain

$$\mathbf{R} = 4\beta^4 u^{\beta-1} t^{2\beta-2} s^{\beta-1} \sum_{1 \le i \le s} i^{-4\beta} \left(1 + \frac{\theta}{i}\right).$$

We compare the quantity on the right with the quantity $\mathbf{H}_{s,t} \mathbf{H}_{t,u}$ evaluated in (28). At this step we use Lemma 1. A straightforward calculation shows that $\mathbf{R} = o(\mathbf{H}_{s,t} \mathbf{H}_{t,u})$ for $2\beta \le 1$. Hence $\mathbf{P}\{\vee_{stu}^0\} = (1 + o(1))\mathbf{H}_{s,t} \mathbf{H}_{t,u}$. Now (28) combined with (23) completes the proof of (38). Proofs of (37) and (40) will be presented in an extended version of the paper.

For $2\beta > 1$ we derive (33) from (36) combined with the bounds

$$\mathbf{P}\{\vee_{stu}^0\} = o(1)\mathbf{P}\{\Delta_{stu}^*\} \quad \text{and} \quad \mathbf{P}\{v_s \sim v_t\}\mathbf{P}\{v_t \sim v_u\} = o(\mathbf{P}\{\Delta_{stu}^*\}).$$

Here the second bound follows by Lemmas 1, 2. To prove the first bound we firstly apply the union bound to (39),

$$\mathbf{P}\{\vee_{stu}^0\} \le \sum_{\bar{w}_1 \in W_s} \sum_{\bar{w}_2 \in W_t \setminus \{\bar{w}_1\}} \mathbf{P}\{\mathcal{A}_{st,\bar{w}_1}\} \mathbf{P}\{\mathcal{A}_{tu,\bar{w}_2}\} \le \mathbf{H}_{s,t} \mathbf{H}_{t,u}.$$

Secondly, we combine (24) with (28) to show that $\mathbf{H}_{s,t} \mathbf{H}_{t,u} = o(\mathbf{P}\{\Delta_{stu}^*\})$. $\square$

Proof of Theorem 1. Let (x^*, y^*, z^*) be an (ordered) triple of distinct integers from $[n]$ sampled uniformly at random. We have

$$C(G_n) = \frac{\mathbf{P}\{\Delta_{x^*y^*z^*}\}}{\mathbf{P}\{\vee_{x^*y^*z^*}\}}. \tag{42}$$

We firstly consider $\mathbf{P}\{\Delta_{x^*y^*z^*}\}$. By symmetry and (32), we have

$$\mathbf{P}\{\Delta_{x^*y^*z^*}\} = \frac{3!}{(n)_3} \sum_{1 \le s < t < u \le n} \mathbf{P}\{\Delta_{stu}\} \tag{43}$$

$$= (1 + o(1))\frac{3!}{(n)_3} \sum_{1 \le s < t < u \le n} \mathbf{P}\{\Delta_{stu}^*\}.$$

We evaluate $\mathbf{P}\{\Delta^*_{stu}\}$ using (24) and then approximate the sum by the integral

$$\sum_{1 \le s < t < u \le n} \mathbf{P}\{\Delta^*_{stu}\}$$

$$= (1 + o(1)) \int_1^n du \int_1^u dt \int_1^t ds \left(\beta^3 \frac{(ut)^{\beta-1}}{s^{2\beta+1}} \left(\mu_s^{(3)} + 3\mu_s^{(2)} + 2\mu_s^{(1)} \right) \right).$$

Note that the functions $s \to \mu_s^{(m)}$, $m = 1, 2, 3$ are given in Lemma 1. A simple calculation shows that the triple integral above (approximately) equals

$$(\beta^3 + o(1)) \times \begin{cases} \frac{c_2}{6\beta^3} n^{3\beta}, & \text{for} \quad \alpha < 2, \\ \frac{6kn \ln n}{(1-2\beta)(1-\beta)}, & \text{for} \quad \alpha = 2, \\ \frac{6kn}{(1-3\beta)(1-2\beta)(1-\beta)}, & \text{for} \quad \alpha > 2. \end{cases}$$

Here and below $c_2 > 0$ is the same as in Lemma 1. Inserting the obtained expression into (43) we get

$$\mathbf{P}\{\Delta_{x^*y^*z^*}\} = (1 + o(1)) \times \begin{cases} c_2 n^{3\beta-3}, & \text{for} \quad \alpha < 2, \\ \frac{36k\beta^3}{(1-2\beta)(1-\beta)} n^{-2} \ln n, & \text{for} \quad \alpha = 2, \quad (44) \\ \frac{36k\beta^3}{(1-3\beta)(1-2\beta)(1-\beta)} n^{-2}, & \text{for} \quad \alpha > 2. \end{cases}$$

We secondly consider $\mathbf{P}\{\vee_{x^*y^*z^*}\} = \mathbf{P}\{v_{x^*} \sim v_{y^*}, v_{y^*} \sim v_{z^*}\}$. By symmetry, we have

$$\mathbf{P}\{\vee_{x^*y^*z^*}\} = \frac{2}{(n)_3} \sum_{1 \le s < t < u \le n} \left(\mathbf{P}\{\vee_{stu}\} + \mathbf{P}\{\vee_{tsu}\} + \mathbf{P}\{\vee_{sut}\} \right). \quad (45)$$

We will evaluate probabilities in the brackets using (33) and Lemmas 1, 2. Then we approximate the sum in (45) by respective (triple) integral as above. We consider the cases $\alpha \le 2$ and $\alpha > 2$ separately.

Let $\alpha \le 2$. For each permutation (x, y, z) of $s < t < u$ we have as $s \to +\infty$ (use (33) and Lemmas 1, 2 and note that $\mathbf{P}\{\vee_{xyz}\} = (1 + o(1))\mathbf{P}\{\Delta^*_{xyz}\}$)

$$\mathbf{P}\{\vee_{xyz}\} = (1 + o(1))\mathbf{P}\{\Delta^*_{stu}\} = (\beta^3 + o(1)) \times \begin{cases} c_2(stu)^{\beta-1}, & \text{for} \quad \alpha < 2, \\ 6k(stu)^{-2/3} \ln s, & \text{for} \quad \alpha = 2. \end{cases}$$

We insert this expression in (45) and then approximate the sum by the integral. We obtain

$$\mathbf{P}\{\vee_{x^*y^*z^*}\} = (1 + o(1)) \times \begin{cases} c_2 n^{3\beta-3}, & \text{for} \quad \alpha < 2, \\ \frac{36k\beta^3}{(1-2\beta)(1-\beta)} n^{-2} \ln n, & \text{for} \quad \alpha = 2. \end{cases} \quad (46)$$

Let $\alpha > 2$. For a permutation (x, y, z) of $s < t < u$ we have as $s \to +\infty$ (use (33) and Lemmas 1, 2)

$$\mathbf{P}\{\vee_{xyz}\} = \frac{(1 + o(1))\beta^3}{u^{1-\beta}t^{1-\beta}} \begin{cases} A(st)^{-\beta} + Bs^{-2\beta}, & \text{for} \quad (x, y, z) = (s, t, u), \\ As^{-2\beta} + Bs^{-2\beta}, & \text{for} \quad (x, y, z) = (t, s, u), \\ Au^{\beta-1}t^{1-2\beta}s^{-\beta} + Bs^{-2\beta}, & \text{for} \quad (x, y, z) = (s, u, t). \end{cases}$$

Here we denote for short $A = \frac{4\beta k^2}{(1-2\beta)^2}$ and $B = \frac{6k}{1-3\beta}$. We insert this expression in (45) and then approximate the sum by the integral. We obtain

$$\mathbf{P}\{\vee_{x^*y^*z^*}\} = \frac{(1+o(1))\beta^3}{n^2}\left(\frac{3A}{(1-\beta)^2} + \frac{2A+6B}{(1-\beta)(1-2\beta)}\right). \qquad (47)$$

Finally, inserting into (42) the approximations of $\mathbf{P}\{\Delta_{x^*y^*z^*}\}$ and $\mathbf{P}\{\vee_{x^*y^*z^*}\}$ given by (44) and (46), (47) we obtain (2). $\square$

Acknowledgement. We thank the anonymous referees for their suggestions and comments.

References

1. Aiello, W., Bonato, A., Cooper, C., Janssen, J., Prałat, P.: A spatial web graph model with local influence regions. Internet Math. **5**, 173–193 (2009)
2. Barabási, A.-L., Albert, R.: Emergence of scaling in random networks. Science **286**(5439), 509–512 (1999)
3. Bloznelis, M.: Degree and clustering coefficient in sparse random intersection graphs. Ann. Appl. Probab. **23**, 1254–1289 (2013)
4. Bloznelis, M., Götze, F.: Preferred attachment model of affiliation network. J. Stat. Phys. **156**, 800–821 (2014)
5. Bloznelis, M., Karoński, M.: Random intersection graph process. In: Bonato, A., Mitzenmacher, M., Prałat, P. (eds.) WAW 2013. LNCS, vol. 8305, pp. 93–105. Springer, Cham (2013). https://doi.org/10.1007/978-3-319-03536-9_8
6. Godehardt, E., Jaworski, J.: Two models of random intersection graphs and their applications. Electron. Notes Discret. Math. **10**, 129–132 (2001)
7. Godehardt, E., Jaworski, J., Rybarczyk, K.: Clustering coefficients of random intersection graphs. In: Gaul, W., Geyer-Schulz, A., SchmidtThieme, L., Kunze, J. (eds.) Challenges at the Interface of Data Analysis, Computer Science, and Optimization. STUDIES CLASS, pp. 243–253. Springer, Heidelberg (2012). https://doi.org/10.1007/978-3-642-24466-7
8. Guillaume, J.L., Latapy, M.: Bipartite structure of all complex networks. Inform. Process. Lett. **90**, 215–221 (2004)
9. Holme, P., Kim, B.J.: Growing scale-free networks with tunable clustering. Phys. Rev. E **65**(2), 026107 (2002)
10. Iskhakov, L., Kamiński, B., Mironov, M., Prałat, P., Prokhorenkova, L.: Local clustering coefficient of spatial preferential attachment model. J. Complex Netw. **8**(1), cnz019 (2020)
11. Jacob, E., Mörters, P.: A spatial preferential attachment model with local clustering. In: Bonato, A., Mitzenmacher, M., Prałat, P. (eds.) WAW 2013. LNCS, vol. 8305, pp. 14–25. Springer, Cham (2013). https://doi.org/10.1007/978-3-319-03536-9_2
12. Lattanzi, S., Sivakumar, D.: Affiliation networks. In: STOC 2009–Proceedings of the: ACM International Symposium on Theory of Computing, pp. 427–434. ACM, New York (2009)
13. Martin, T., Ball, B., Karrer, B., Newman, M.E.J.: Coauthorship and citation patterns in the Physical Review. Phys. Rev. E **88**, 012814 (2013)

14. Newman, M.E.J., Watts, D.J., Strogatz, S.H.: Random graph models of social networks. Proc. Natl. Acad. Sci. U.S.A. **99**(Suppl. 1), 2566–2572 (2002)
15. Ostroumova, L., Ryabchenko, A., Samosvat, E.: Generalized preferential attachment: tunable power-law degree distribution and clustering coefficient. In: Bonato, A., Mitzenmacher, M., Prałat, P. (eds.) WAW 2013. LNCS, vol. 8305, pp. 185–202. Springer, Cham (2013). https://doi.org/10.1007/978-3-319-03536-9_15

Transience Versus Recurrence
for Scale-Free Spatial Networks

Peter Gracar[1]([✉])[iD], Markus Heydenreich[2][iD], Christian Mönch[3][iD],
and Peter Mörters[1][iD]

[1] Mathematisches Institut, Universität zu Köln, Weyertal 86-90,
50931 Cologne, Germany
{pgracar,moerters}@math.uni-koeln.de
[2] Mathematisches Institut, Ludwig-Maximilians-Universität München,
Theresienstr. 39, 80333 Munich, Germany
m.heydenreich@lmu.de
[3] Institut für Mathematik, Johannes Gutenberg-Universität Mainz,
Staudingerweg 9, 55099 Mainz, Germany
cmoench@uni-mainz.de

Abstract. Weight-dependent random connection graphs are a class of
local network models that combine scale-free degree distribution, small-
world properties and clustering. In this paper we discuss recurrence or
transience of these graphs, features that are relevant for the performance
of search and information diffusion algorithms on the network.

Keywords: Random-connection model · Recurrence · Transience ·
Scale-free percolation · Preferential attachment · Boolean model

1 Introduction and Statement of Results

1.1 Motivation

In the age of "big data" we are increasingly faced with data that is not linearly
structured and instead organised in the form of networks. Algorithmic processing
of such data is often dependent on the topological connectivity properties of the
network. In this paper we therefore investigate finer connectivity features of a
range of random network models. Features shared by these models are:

- They are *scale-free*, i.e., the proportion of nodes with k neighbours is of order
  $k^{-\tau+o(1)}$ for some *power law exponent* τ.
- They are *clustering*, i.e., local neighbourhoods of a node have a much higher
  connectivity than the overall network.
- They are *small worlds*, i.e., the graph distances are no more than polyloga-
  rithmic with respect to the system size.

Under the further assumption that the power law exponent τ is sufficiently small,
the models have the following additional features:

© Springer Nature Switzerland AG 2020
B. Kamiński et al. (Eds.): WAW 2020, LNCS 12091, pp. 96–110, 2020.
https://doi.org/10.1007/978-3-030-48478-1_7

- They are *ultrasmall*, i.e., the shortest path between two typical nodes in the graph is doubly logarithmic in the size of the graph.
- They are *robust*, i.e., if an arbitrarily large proportion of links is randomly removed the qualitative features of the network remain unchanged.

A prototype of such a network is the *age-based spatial preferential attachment model* introduced in [8]. In this model nodes arrive after exponential waiting times and upon birth are placed randomly on the unit torus $\mathbb{T}^d$. They connect independently to all existing nodes with a probability which is a decreasing function of the spatial distance and the birth times of both vertices. This network model is a simplification of the spatial preferential attachment model in [16] and, in a less general setup, in [1], which however is believed to retain all essential features of the more complex original spatial preferential attachment model.

The major tool to study the age-based spatial preferential attachment model is to look at a local limit graph on $\mathbb{R}^d$. Such a graph describes the scaled neighbourhoods of a typical vertex in the network at a large time; long term averaged features of the network are reflected in the features of the limiting graph. It is shown in [8] that the limiting graph for the age-based spatial preferential attachment model is the *age-dependent random connection model*, which is a special case of the class of weight-dependent random connection models studied in this paper and introduced below. This connection is used in [8] to show scale-freeness and clustering, and in [9,11] to identify regimes of robustness and ultrasmallness of the age-based spatial preferential attachment model.

In the context of information propagation, further properties of networks are relevant: How long does it take for the information to propagate and reach a set of targets for the first time? Does a target node receive the information at all (or is it possible that information bypasses it)? Can a single source result in information reaching a target in more than one way? Conversely, can information travel through the network without being "detected" by a large proportion of the network? Such questions are tightly connected to (and often well described by) the behaviour of a random walker on the network. In many cases, the behaviour of the walker is crucial for the development of random walk based search strategies, cf. [27]. The present paper addresses the problem of recurrence or transience of the limiting graph. Recurrence means that a random walker returns infinitely often to its origin and it is a prerequisite for the functioning of many search and information diffusion algorithms on networks [23].

1.2   The Weight-Dependent Random Connection Model

We study the transient/recurrent behaviour of a class of infinite graphs that, although not necessarily built as a limit of a growing sequence of finite graphs as in our motivating example, are built using similarly simple rules to connect pairs of vertices. We call this class of graphs the *weight-dependent random connection model* and we now describe the building principles they all have in common.

The vertex set of the model is a Poisson process of unit intensity on $\mathbb{R}^d \times [0,1]$. We think of a Poisson point $\mathbf{x} = (x,s)$ as a *vertex* at *position* x with *weight* s^{-1}.

Two vertices $\mathbf{x} = (x, s)$ and $\mathbf{y} = (y, t)$ are connected by an edge with probability $\varphi(\mathbf{x}, \mathbf{y})$ for a connectivity function

$$\varphi \colon (\mathbb{R}^d \times [0, 1]) \times (\mathbb{R}^d \times [0, 1]) \to [0, 1], \tag{1}$$

satisfying $\varphi(\mathbf{x}, \mathbf{y}) = \varphi(\mathbf{y}, \mathbf{x})$. Connections between different (unordered) pairs of vertices occur independently. We assume throughout that φ has the form

$$\varphi(\mathbf{x}, \mathbf{y}) = \varphi\big((x, s), (y, t)\big) = \rho\big(\beta^{-1} g(s, t) |x - y|^d\big) \tag{2}$$

for a non-increasing, integrable *profile function* $\rho \colon \mathbb{R}_+ \to [0, 1]$, a *percolation parameter* $\beta > 0$ and a *kernel function* $g \colon [0, 1] \times [0, 1] \to \mathbb{R}_+$. The parameter β controls the edge density in a monotone way; increasing β increases the number of edges connected to a vertex at the origin. Varying β can also be interpreted as rescaling distances, and therefore it is equivalent to varying the intensity of the underlying Poisson process. We assume further that g is non-increasing in both arguments; and ρ is non-increasing, so that vertices whose positions are far apart are less likely to be connected. Without loss of generality we scale the profile function as

$$\int_{\mathbb{R}^d} \rho(|x|^d) \, dx = 1. \tag{3}$$

Then it is easy to see that the degree distribution of a vertex does not depend on ρ (see for example [8, Proposition 4.1]); it does however influence the likelihood of long edges.

We next give concrete examples for the kernel function g, and demonstrate that our setup yields a number of well-known models in continuum percolation theory. We define the functions in terms of a parameter $\gamma \in (0, 1)$, which determines the strength of the influence of the vertex weight on the connection probabilities; large γ correspond to strong favouring of vertices with large weight. The models considered below are all scale-free with power-law exponent

$$\tau = 1 + \frac{1}{\gamma},$$

which means that, in probability as $N \to \infty$,

$$\frac{\#\text{ vertices } x \in [-N, N]^d \text{ with degree k}}{\#\text{ vertices } x \in [-N, N]^d} \to \mu(k), \text{ and } \mu(k) = k^{-\tau + o(1)}. \tag{4}$$

(A) We define the *plain kernel* as

$$g^{\text{plain}}(s, t) = 1. \tag{5}$$

In this case we have no dependence on the weights. If $\rho(r) = 1_{[0, a]}$ for $a = d/\omega_d$ and ω_d is the area of the unit sphere in $\mathbb{R}^d$, this gives the *Gilbert disc model* with radius $\sqrt[d]{\beta a}$. Functions ρ of more general form lead to the (ordinary) *random connection model*, including in particular a continuum version of *long-range percolation* when ρ has polynomial decay at infinity.

(B) We define the *sum kernel* as

$$g^{\mathrm{sum}}(s,t) = \left(s^{-\gamma/d} + t^{-\gamma/d}\right)^{-d}. \tag{6}$$

Interpreting $(as^{-\gamma})^{1/d}, (at^{-\gamma})^{1/d}$ as random radii and letting $\rho(r) = 1_{[0,a]}$ we get the *Boolean model* in which two vertices are connected by an edge if the associated balls intersect. We get a further variant of the model with the *min-kernel* defined as

$$g^{\mathrm{min}}(s,t) = (s \wedge t)^{\gamma},$$

which, as $g^{\mathrm{sum}} \leq g^{\mathrm{min}} \leq 2g^{\mathrm{sum}}$, shows qualitatively the same behaviour.
(C) For the *max-kernel* defined as

$$g^{\mathrm{max}}(s,t) = (s \vee t)^{1+\gamma},$$

we may choose any $\gamma > 0$. This is a continuum version and generalization of the ultra-small scale-free geometric networks of Yukich [28], which is also parametrized to have power-law exponent $\tau = 1 + \frac{1}{\gamma}$.
(D) A particularly interesting case is the *product kernel*

$$g^{\mathrm{prod}}(s,t) = s^{\gamma}t^{\gamma}, \tag{7}$$

which leads to a continuum version of the *scale-free percolation* model of Deijfen et al. [4,13], see also [5,6]. This model combines features of scale-free random graphs and polynomial-decay long-range percolation models (for suitable choice of ρ).
(E) Our final example for g is the *preferential attachment kernel*

$$g^{\mathrm{pa}}(s,t) = (s \vee t)^{1-\gamma}(s \wedge t)^{\gamma}, \tag{8}$$

which gives rise to the *age-dependent random connection model* introduced by Gracar et al. [8] as local weak limit of the age-based preferential attachment model, which is a simplification and approximation of the spatial preferential attachment model in Jacob and Mörters [16]. In this model, s and t actually play the role of the birth times of vertices in the underlying dynamic network. This model also combines scale-free degree distributions with power-law exponent $\tau = 1 + \frac{1}{\gamma}$ and long edges in a natural way, but has a fundamentally different graph topology, as we will see.

The above listed kernels represent only some of the possible choices. We refer the reader to Table 1 for a short literature survey of the terminology under which the above kernels appear in the literature.

1.3 Main Results

We now focus on a profile function with polynomial decay

$$\lim_{v \to \infty} \rho(v)\, v^{\delta} = 1 \quad \text{for a parameter } \delta > 1, \tag{9}$$

Table 1. Terminology of the models in the literature.

Vertices	Profile	Kernel	Name and reference
Poisson	Indicator	Plain	Random geometric graph, Gilbert disc model [24]
Poisson	General	Plain	Random connection model [20]
			Soft random geometric graph [25]
Lattice	Polynomial	Plain	Long-range percolation [2]
Poisson	Indicator	Sum	Boolean model [12,21]
Lattice	Indicator	Max	Ultra-small scale-free geometric networks [28]
Poisson	Indicator	Min	Scale-free Gilbert graph [15]
Lattice	Polynomial	Prod	Scale-free percolation [4,13]
Poisson	Polynomial	Prod	Inhomogeneous long-range percolation [5]
			Continuum scale-free percolation [6]
Poisson	General	Prod	Geometric inhomogeneous random graphs [3]
Poisson	General	Pa	Age-dependent random connection model [8]

renormalised appropriately to satisfy (3), and fix one of the kernel functions described above. We keep γ, δ fixed and study the resulting graph $\mathcal{G}^\beta$ as a function of β. Note that our assumptions $\delta > 1$ and $\gamma < 1$ guarantee that $\mathcal{G}^\beta$ is locally finite for all values of β, since it can quickly be checked that the degree of every vertex is Poisson distributed, cf. [8, p. 315 and Proposition 4.1]. We informally define β_c as the infimum over all values of β such that $\mathcal{G}^\beta$ contains an infinite component (henceforth the infinite *cluster*). If $d \geq 2$, we always have $\beta_c < \infty$, cf. [13]. General arguments in [7] yield that there is at most one infinite component of $\mathcal{G}^\beta$, and hence there is a unique infinite component whenever $\beta > \beta_c$. We study the properties of this infinite cluster.

Two cases correspond to different network topologies.

- If $\gamma > \frac{1}{2}$ for the product kernel, $\gamma > 0$ for the max kernel, or $\gamma > \frac{\delta}{\delta+1}$ for the preferential attachment, min or sum kernels we have $\beta_c = 0$, i.e. there exists an infinite cluster irrespective of the edge density, see [13] for product, [28] for max and [11,17] for all other kernels. We say that this is the *robust case*.
- Otherwise, if $\gamma < \frac{1}{2}$ for the product kernel, see [13], or if $\gamma < \frac{\delta}{\delta+1}$ for the preferential attachment, min or sum kernels, we have $\beta_c > 0$. This was recently shown in [11]. In this case we say we are in the *non-robust case*.

Our main interest is whether the infinite cluster is recurrent (i.e., simple random walk on the cluster returns to its starting point almost surely), or transient (i.e., simple random walk on the cluster never returns to its starting point with positive probability). Our results are summarized in the following theorem.

Theorem 1 (Recurrence vs. transience of the weight-dependent random connection model). *Consider the weight-dependent random connection model with profile function satisfying (9).*

(a) *For preferential attachment kernel, sum kernel, or min kernel and $\beta > \beta_c$,*
    the infinite component is
    – transient *if either* $1 < \delta < 2$ *or* $\gamma > \delta/(\delta + 1)$;
    – recurrent *if* $d \in \{1, 2\}$, $\delta > 2$ *and if* $\gamma < \delta/(\delta + 1)$.
(b) *For the product kernel and $\beta > \beta_c$, the infinite component is*
    – transient *if either* $1 < \delta < 2$ *or* $\gamma > 1/2$;
    – recurrent *if* $d \in \{1, 2\}$, $\delta > 2$ *and if* $\gamma < 1/2$.
(c) *For the max kernel and $\beta > \beta_c$, the infinite component is* transient.

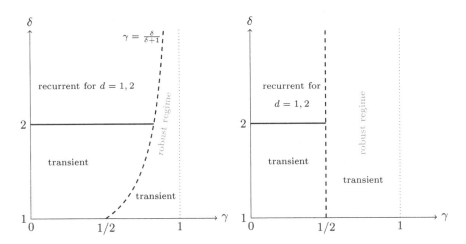

Fig. 1. The different phases in Theorem 1: Left: preferential attachment kernel. Right: product kernel. The dashed line separates the robust from the non-robust phase.

For a summary of the results we refer to Fig. 1. We describe the proof of this theorem in Sects. 3 and 4, and refer to the journal version [10] for the full argument. We emphasise again that the characterisation of the regimes for this large class of graphs yields important information about them, useful for example in order to determine whether random walker algorithms will be able to properly scale to larger graphs or not.

Remarks:

– Loosely speaking, for the models in (a) and (b) the walk can travel to infinity using long edges if there are enough of them, i.e. if $\delta < 2$. For the models in (a) the walk can also use that vertices of ever increasing weight can be reached using a pool of intermediate vertices, which is big enough if $\gamma > \delta/(\delta + 1)$. For the model in (b) with $\gamma > \frac{1}{2}$ and the model in (c) with $0 < \gamma < 1$ the walk can travel directly between vertices of increasing weight without using intermediate edges.

– When $\delta > 2$ and $\gamma < \frac{\delta}{\delta+1}$ for the preferential attachment kernel resp. $\gamma < \frac{1}{2}$ for the product kernel, we expect that the long-range and scale-free effects do not influence the behaviour of the random walk, so that for $d \geq 3$ the infinite cluster would be transient. A mathematical proof of transience in this regime (even for long-range percolation) has not yet been found. We plan to address this in future work.

2   The Weight-Dependent Random Connection Model

Construction As a Point Process on $(\mathbb{R}^d \times [0,1])^{[2]} \times [0,1]$. We give now a more formal construction of our model. To this end, we extend the construction given in [14, Sections 2.1 and 2.2] by additional vertex marks (the *weight* or birth time). For further constructions, see Last and Ziesche [19] and Meester and Roy [22]. We construct the random connection model as a deterministic functional $\mathcal{G}_\varphi(\xi)$ of a suitable point process ξ. Let η denote a unit intensity $\mathbb{R}^d$-valued Poisson point process, which we can write as

$$\eta = \{X_i \colon i \in \mathbb{N}\}; \tag{10}$$

such enumeration is possible, cf. [18, Corollary 6.5]. In order to define random walks on the random connection model, it is convenient to have a designated (starting) point, and we therefore add an extra point $X_0 = 0$ and thereby get a Palm version of the Poisson point process.

We further equip any Poisson point X_i $(i \geq 0)$ with an independent mark S_i drawn uniformly from the interval $(0,1)$. This defines a point process $\eta' := \{\mathbf{X}_i = (X_i, S_i) \colon i \in \mathbb{N}_0\}$ on $\mathbb{R}^d \times (0,1)$. Let $(\mathbb{R}^d \times (0,1))^{[2]}$ denote the space of all sets $e \subset \mathbb{R}^d \times (0,1)$ with exactly two elements; these are the potential edges of the graph. We further introduce independent random variables $(U_{i,j} : i, j \in \mathbb{N}_0)$ uniformly distributed on the unit interval $(0,1)$ such that the double sequence $(U_{i,j})$ is independent of η'. Using $<$ for the strict lexicographical order on $\mathbb{R}^d$, we can now define

$$\xi := \left\{ \left(\{(X_i, S_i), (X_j, S_j)\}, U_{i,j} \right) : X_i < X_j, i, j \in \mathbb{N}_0 \right\}, \tag{11}$$

which is a point process on $(\mathbb{R}^d \times (0,1))^{[2]} \times (0,1)$. Mind that η' might be recovered from ξ. Even though the definition of ξ does depend on the ordering of the points of η, its distribution does not.

We can now define the weight-dependent random connection model $\mathcal{G}_\varphi(\xi)$ as a deterministic functional of ξ; its vertex and edge sets are given as

$$V(\mathcal{G}_\varphi(\xi)) = \eta',$$
$$E(\mathcal{G}_\varphi(\xi)) = \{\{\mathbf{X}_i, \mathbf{X}_j\} \in V(\mathcal{G}_\varphi(\xi))^{[2]} : X_i < X_j, U_{i,j} \leq \varphi(\mathbf{X}_i, \mathbf{X}_j), i, j \in \mathbb{N}_0\}.$$

Only in this section we write $\mathcal{G}_\varphi(\xi)$ in order to make the dependence on the connection function φ explicit; in the following sections we will fix a kernel function as well as the parameters δ and γ, and therefore only write $\mathcal{G}^\beta = \mathcal{G}^\beta(\xi)$.

Percolation. Our construction ensures that $\mathbf{0} := (X_0, S_0) \in V(\mathcal{G}_\varphi(\xi))$. We now write $\{0 \leftrightarrow \infty\}$ for the event that the random graph $\mathcal{G}_\varphi(\xi)$ contains an infinite self-avoiding path $(v_1, v_2, v_3, \dots)$ of vertices with $v_i \in V(\mathcal{G}_\varphi(\xi))$ ($i \in \mathbb{N}$) such that $\{\mathbf{0}, v_1\}, \{v_1, v_2\}, \{v_2, v_3\} \cdots \in E(\mathcal{G}_\varphi(\xi))$, and we say that in this case the graph *percolates*. We denote the percolation probability by

$$\theta(\beta) = \mathbb{P}(0 \leftrightarrow \infty \text{ in } \mathcal{G}_\varphi(\xi));$$

for the probability that this happens; this quantity is well-defined by the monotonicity of the right-hand side in β. This allows us to define the critical percolation threshold as

$$\beta_c := \inf\{\beta > 0 : \theta(\beta) > 0\} \geq 0. \tag{12}$$

Random Walks. We recall that, as $\gamma < 1$, the resulting graph $\mathcal{G}_\varphi(\xi)$ is locally finite, i.e.

$$\sum_{\mathbf{y} \in V(\mathcal{G}_\varphi(\xi))} \mathbb{1}\{\{\mathbf{x}, \mathbf{y}\} \in E(\mathcal{G}_\varphi(\xi))\} < \infty \qquad \text{for all } \mathbf{x} \in V(\mathcal{G}_\varphi(\xi)) \text{ almost surely,}$$

$$\tag{13}$$

cf. [8, p. 315 and Proposition 4.1]. Given $\mathcal{G}_\varphi(\xi)$ with $0 \leftrightarrow \infty$ we define the *simple random walk* on the random graph $\mathcal{G}_\varphi(\xi)$ as the discrete-time stochastic process for which $X_0 = 0$ and

$$P^{\mathcal{G}_\varphi(\xi)}(X_n = y \mid X_{n-1} = x) = \frac{\mathbb{1}\{\{\mathbf{x}, \mathbf{y}\} \in E(\mathcal{G}_\varphi(\xi))\}}{\sum_{\mathbf{z} \in V(\mathcal{G}_\varphi(\xi))} \mathbb{1}\{\{\mathbf{x}, \mathbf{z}\} \in E(\mathcal{G}_\varphi(\xi))\}}$$

for any $\mathbf{x}, \mathbf{y} \in V(\mathcal{G}_\varphi(\xi))$ and $n \in \mathbb{N}$. We say that $\mathcal{G}_\varphi(\xi)$ is *recurrent* if

$$P^{\mathcal{G}_\varphi(\xi)}\big(\exists n \geq 1 : X_n = 0\big) = 1,$$

otherwise we say that it is *transient*.

3   Transience

In this section we outline the key steps needed to prove the transience statement of Theorem 1. Throughout, we write $\mathcal{G}^\beta$ instead of $\mathcal{G}_\varphi(\xi)$ to stress that kernel and profile are fixed and the percolation parameter is β.

3.1   Transience in the Robust Case

Proving transient behaviour for the robust case hinges on a renormalisation sequence argument. Heuristically, we consider a finite box in $\mathbb{R}^d$ and the largest cluster of connected vertices inside of this box. Then, if the box is chosen sufficiently large, the probability that this cluster represents a positive proportion of the entire vertex set is increasing, as is the probability that within this cluster a vertex with weight greater than some predetermined value exists. When these two conditions are satisfied, we consider this box good. We now, roughly speaking, repeat this argument for a considerably larger box. We break this large box into disjoint boxes of the previous scale and consider only those boxes which satisfy the two conditions (which occurs independently and with uniform probability for all boxes). Then, we call the bigger box good whenever a sufficiently large proportion of the boxes contained therein are good, they are sufficiently well connected with each other and there exists a vertex in this newly constructed cluster with weight greater than an even larger predetermined value.

Repeating this procedure at greater and greater scales we obtain a renormalised graph sequence that is contained in the infinite component of the graph and can be shown to be transient with a fairly straightforward argument. We formalise this statement in the following two results and leave the proof of Theorem 2 for the full version of this paper [10].

Given a graph $G = (V, E)$ and a sequence $\{C_n\}_{n=1}^{\infty}$ let $V_l(j_l, \ldots, j_1)$ with $l \in \mathbb{N}_0$ and $j_n \in \{1, \ldots, C_n\}$ be an element of the vertex set V. Furthermore, let V_0 be some arbitrary vertex. Now let for $l \geq m > 1$

$$V_l(j_l, \ldots, j_m) = \bigcup_{j_{m-1}=1}^{C_{m-1}} \cdots \bigcup_{j_1=1}^{C_1} V_l(j_l, \ldots, j_1).$$

We call the sets $V_l(j_l, \ldots, j_m)$ *bags*, and the numbers C_n *bag sizes*.

Definition 1. *We say that the graph* $G = (V, E)$ *is renormalized for the sequence* $\{C_n\}_{n=1}^{\infty}$ *if we can construct an infinite sequence of graphs such that the vertices of the l-th stage graph are labelled by* $V_l(j_l, \ldots, j_1)$ *for all* $j_n \in \{1, \ldots, C_n\}$, *and such that for every* $l \geq m > 2$, *every* $j_l, \ldots, j_{m+1}$, *and all pairs of distinct* $u_m, w_m \in \{1, \ldots, C_m\}$ *and* $u_{m-1}, w_{m-1} \in \{1, \ldots, C_{m-1}\}$ *there is an edge in* G *between a vertex in the bag* $V_l(j_l, \ldots, j_{m+1}, u_m, u_{m-1})$ *and a vertex in the bag* $V_l(j_l, \ldots, j_{m+1}, w_m, w_{m-1})$.

The underlying intuition is that every n-th stage bag contains C_n $(n-1)$-stage bags, each one of which contains again C_{n-1} $(n-2)$-stage bags. Every pair of $(n-2)$-stage bags in an n stage bag is connected by an edge between one of the vertices in the bags.

Lemma 1 (Berger, [2, Lemma 2.7]). *A graph renormalized for the sequence* C_n *is transient if* $\sum_{n=1}^{\infty} C_n^{-1} < \infty$.

Theorem 2 (Product and max kernel contain a renormalised graph sequence). *Let* $\beta > \beta_c$ *for the chosen kernel. If* $\gamma > \frac{1}{2}$ *for* g^{prod} *or* $\gamma > 0$ *for* g^{max}, *then the infinite connected component contains a graph renormalized for the sequence* $C_n = (n+1)^{2d}$ *almost surely. Consequently, the infinite component is transient.*

Before we proceed to argue the result for the remaining kernels from Theorem 1, we first give a heuristic argument as to why the proof for the max and product kernel does not work for the rest. In order to demonstrate the argument, we assume for the moment that the profile function ρ is the indicator function $\mathbb{1}_{[0,c]}$ for some constant c. The first key observation is that two nodes of the graph that are far apart can only be connected if both of their respective weights are large. This is especially clear in the max kernel case, since a connection between two nodes at distance v is then only possible when *both* of their weights are greater than $(v^d/(\beta c))^{1/(1+\gamma)}$. A similar observation can be made for the product kernel - if one of the weights is small, the other weight must be considerably bigger to ensure their product is large enough.

For the other kernels, both weights being large is similarly beneficial; it should be remarked however that the probability of both weights being large is not sufficiently offset by the increase in the connection probability. Unlike in the above example, only the first heavy weight leads to a big increase in the probability of a connection existing (the second weight has a considerably smaller effect on the probability). In contrast to the max and product kernel, however, if the profile function has sufficiently heavy tails at infinity, an alternative strategy exists. We can connect pairs of nodes with large weights through a connector node with a comparatively small weight. Intuitively, since the smaller of the weights in a pair does not affect the connection probability of two nodes for the min kernel (and affects the probability at a lower order than the large weight in the sum and pa kernel), we can attempt to connect two far away nodes with large weights through nodes of smaller weight. Their large number then makes the probability of successfully finding such two-step connection sufficiently high to again obtain a renormalised graph sequence. It is helpful to interpret this event (i.e., the existence of a connector node through with which both nodes of the pair are connected) as the existence of a *bridged connection* in a new graph with the same vertex set (Fig. 2). Note however that a connection in this new graph does not correspond to a specific tuple of a connector node and two corresponding edges; instead, a bridged connection corresponds to the *existence* of such a tuple. Then, one can roughly speaking use the same strategy as before in this new graph.

Therefore, although the construction of Theorem 2 does not yield the desired renormalised graph sequence for the remaining kernels when using direct connections, it does lead to the stated result when using the bridged connections instead. We state the main properties that hold for these connections in the following proposition.

Proposition 1 (Occurrence of bridged connections). *Let (x,t) and (y,s) be two nodes of the graph with $t > s$. Then there exists a positive constant C such that the probability that there exists a bridged connection between the two vertices is at least*

$$1 - \exp\{-Cs^{-\gamma}\rho(\beta^{-1}t^{\gamma}(s^{\frac{-\gamma}{d}} + |x - y|)^d)\}.$$

Furthermore, this probability is monotonically decreasing in $|x - y|$.

A direct consequence of Proposition 1 is that one can, using the same construction as used to prove Theorem 2, obtain a renormalised graph sequence satisfying the conditions of Lemma 1 which leads to the following result.

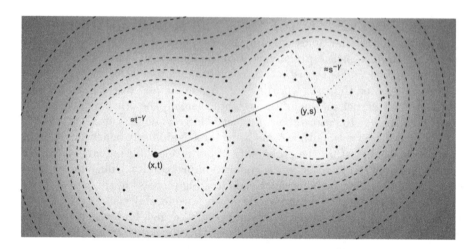

Fig. 2. The diagram illustrates the intensity of the Poisson point process of all points that are connected to (x,t) and (y,s) for the pa kernel and a large δ. Intuitively, the area where such points are probable (contained roughly within the 4th innermost contour line) is sufficiently large to make the existence of a *bridged connection* (like the one in red) more probable than the direct connection (which would fall well outside the 4th innermost contour line of only (x,t) or (y,t)).

Theorem 3 (Min, sum and preferential attachment kernel contain a renormalised graph sequence). *Suppose $\gamma > \frac{\delta}{\delta+1}$ and the kernel is g^{min}, g^{sum} or g^{pa}. Then if $\beta > \beta_c$ the infinite connected component contains a graph renormalized for the sequence $C_n = (n+1)^{2d}$ almost surely. Consequently, the infinite component is transient.*

3.2   Transience in the Non-robust Case

Similar to the robust case, transience in the non-robust case requires a renormalisation argument. However, unlike in the previous section, the paths which carry the random walk out of any finite neighbourhood of $\mathbf{0}$ are not supported by vertices of extremely large weight which in turn are incident to very long edges, independently of the overall density of edges. Instead, the walk travels along a multitude of moderately long edges; if $\delta < 2$ and $\beta > \beta_c$ then there are sufficiently many of these edges and the walk is transient. The reason behind this difference is the same structural feature that distinguishes robustness from non-robustness: In the robust phase, there is a backbone of very few hubs of extremely high weight that guarantees a high connectivity of the network, whereas in the non-robust phase, these hubs are absent and high connectivity can only be obtained by strongly amplifying the edge density.

We have seen in Sect. 3.1 that the precise way of forming the connections between vertices of large weight in the robust case depends on the form of the kernel function g. In the non-robust case, the form of the kernel g is much less important for the proof of transience, and only requires that the profile decays sufficiently slowly. More precisely, we require only the existence of s_* such that

$$\liminf_{v \to \infty} \rho\big(g(s_*, s_*) v^d\big) v^{\delta d} > 0, \tag{14}$$

for some $\delta < 2$. As soon as (14) is satisfied, any supercritical weight-dependent random connection model contains a sub-graph that shows the same qualitative behaviour as a supercritical cluster in long-range percolation with tail exponent δ, which is known to be transient [2].

Theorem 4 (Non-robust supercritical clusters are transient if $\delta < 2$). *Let $\mathcal{G} = \mathcal{G}_\varphi(\xi)$ denote the weight-dependent random connection model with ρ, g satisfying (14) for some $\delta < 2$. If $\mathcal{G}$ is supercritical, then the infinite cluster is transient.*

For a detailed proof of Theorem 4 we refer the reader to the full version of this paper [10]. Here, we provide a condensed version of our argument and briefly discuss its limitations. To relate $\mathcal{G}$ to long-range percolation, we use a coarse graining technique. $\mathbb{R}^d$ is partitioned into cubes and these cubes form the sites in a *long-range bond-site percolation model*. Connectivity *between* sites is established using the edge-probabilities inherited from the underlying weight-dependent random connection model. The crucial ingredient is that, for a site to be *present at all* in the coarse grained model, the corresponding cube needs to contain a cluster of $\mathcal{G}$ that is sufficiently dense in the cube.

Proposition 2 (Local density of clusters). *Let $\mathcal{G}$ be as in Theorem 4. For any $\lambda \in (0,1)$, and any $\varepsilon > 0$, there is a sufficiently large $M_0 \in \mathbb{N}$, such that the following is true for all $M > M_0$: the probability that the cube $[0, M)^d$ contains a cluster with at least $M^{\lambda d}$ vertices exceeds $1 - \varepsilon$.*

The proof of Proposition 2 is precisely where the renormalisation scheme mentioned above comes into play. Just as in the robust case, vertices are grouped into boxes on the initial scale. Boxes are called *good* if the vertices inside a box form sufficiently large clusters, and *bad* if this is not the case. On all subsequent scales, boxes themselves are grouped into larger boxes. The larger boxes are in turn good if they have many good sub-boxes and the clusters inside these good sub-boxes are sufficiently well-connected with each other, and thus form a single cluster on a larger scale. When proceeding upwards in this hierarchy, one needs to control the probability that several sub-clusters inside a box do not belong to a single larger cluster. Our estimate for this probability is obtained from an auxiliary construction describing how clusters inside boxes merge. To formulate it, let $\delta_0 \in (1,2)$, $\mathbf{m} = (m_1, \ldots, m_r)$ with $r \in \mathbb{N}$ and $m_j \in \mathbb{N}, j = 1, \ldots, r$ and let $\mathcal{I}_{\mathbf{m}, \delta_0}$ denote the *inhomogeneous random graph* which is constructed on the vertex set $\{1, \ldots, r\}$ by creating edges between $1 \le i < j \le r$ independently with probability

$$1 - e^{-m_i m_j / (\Sigma_{k=1}^r m_k)^{\delta_0}}.$$

It is instructive to interpret $\mathbf{m}$ as mass distribution and $|\mathbf{m}| := \sum_{k=1}^r m_k$ as *total mass* of $\mathcal{I}_{\mathbf{m}, \delta_0}$. At any stage of the renormalisation scheme, the clusters inherited from the previous stage are interpreted as the vertices in an inhomogeneous random graph and the cluster sizes as the corresponding masses. Our probability bounds for the merger of clusters rely on the following result:

Lemma 2 (Berger, [2, Lemma 2.5]). *Let $\delta_0 \in (1, 2)$ and $\varrho \in (0, 1)$ such that*

$$18\varrho > 16 + \delta_0.$$

There exist $\zeta = \zeta(\delta_0, \varrho) > 0$ and $M_0(\delta_0, \varrho) > 0$ such that for all $\mathbf{m}$ with $|\mathbf{m}| \geq M_0$

$$\mathbb{P}\left(N_{|\mathbf{m}|^\varrho}(\mathcal{I}_{\mathbf{m}, \delta_0}) > 1\right) < |\mathbf{m}|^{-\zeta},$$

where $N_x(\mathcal{I}_{\mathbf{m}, \delta_0})$ denotes the number of clusters $C \subset \mathcal{I}_{\mathbf{m}, \delta}$ with $\sum_{j \in C} m_j \geq x$.

Note that the assumption that $\delta \leq \delta_0 < 2$ is necessary to apply Lemma 2. This is precisely the reason why neither for the weight-dependent random connection model, nor for any other known long-range percolation model with polynomial tail connection probabilities, the proof of transience in the non-robust case can be extended to the case where $\delta \geq 2$.

After invoking Lemma 2, a union bound over all stages of the renormalisation shows that if the scaling parameters are carefully tuned, then the total probability of ever encountering a bad box when zooming outward from $\mathbf{0}$ can be made arbitrarily small, which implies Proposition 2. In turn, Proposition 2 implies that the site density in the coarse grained model can be brought arbitrarily close to one and thus the transience of the coarse-grained long-range bond-site percolation model follows from the corresponding result for the long-range bond percolation model.

4   Recurrence

In order to show recurrence in dimensions $d \in \{1, 2\}$, we use electrical network theory.

Proposition 3 (Nash-Williams, [26]). *Let G be a graph with conductance C_e on every edge e. Consider a random walk on the graph such that when the particle is at some vertex, it chooses its way with probabilities proportional to the conductances on the edges that it sees. Let $\{\Pi_n\}_{n=1}^\infty$ be disjoint cut-sets, and denote by C_{Π_n} the sum of the conductances of Π_n. If*

$$\sum_n C_{\Pi_n}^{-1} = \infty,$$

then the random walk is recurrent.

The arguments that lead to the result for $d = 1$ and $d = 2$ are subtly different from each other, but roughly correspond to showing that the number of edges leaving disjoint cut-sets are sufficiently light tailed. In dimension 1, this can be shown directly by simply treating each edge as having unit conductance and counting the expected number of edges connecting subsequent cut-sets. In dimension 2, we consider instead a lattice based graph that is constructed so as to have effective conductance that is not smaller than that of our random graph. Then, using a projection argument similar to the one in [2] on this lattice graph, a sufficient condition for recurrence can again be obtained. Putting these arguments together, we obtain the following result.

Theorem 5 (Recurrence in one and two dimensions, [2]). *For $d = 1$ let G be a random graph on a unit intensity Poisson point process where two vertices x and y are connected with probability $P_{|x-y|}$ such that*

$$\limsup_{v \to \infty} v^2 P_v < \infty.$$

For $d = 2$ let G be a random graph on a unit intensity Poisson point process where two vertices x and y are connected with probability $P_{|x_1-y_1|,|x_2-y_2|}$ such that

$$\limsup_{u,v \to \infty} (u + v)^4 P_{u,v} < \infty.$$

In both cases, any infinite component of such graph is recurrent.

Consequently in dimensions 1 and 2, if $\delta > 2$ and $\gamma < \delta/(\delta + 1)$ for the preferential attachment, sum and min kernels, any infinite component is recurrent. Similarly, if $\delta > 2$ and $\gamma < 1/2$ for the product kernel, any infinite component is recurrent.

Acknowledgement. We acknowledge support from DFG through Scientific Network *Stochastic Processes on Evolving Networks.*

References

1. Aiello, W., Bonato, A., Cooper, C., Janssen, J., Prałat, P.: A spatial web graph model with local influence regions. Internet Math. **5**(1–2), 175–196 (2008)
2. Berger, N.: Transience, recurrence and critical behavior for long-range percolation. Commun. Math. Phys. **226**(3), 531–558 (2002). Corrected proof of Lemma 2.3 at arXiv:math/0110296v3
3. Bringmann, K., Keusch, R., Lengler, J.: Geometric inhomogeneous random graphs. Theoret. Comput. Sci. **760**, 35–54 (2019)
4. Deijfen, M., van der Hofstad, R., Hooghiemstra, G.: Scale-free percolation. Ann. Inst. Henri Poincaré Probab. Stat. **49**(3), 817–838 (2013)
5. Deprez, P., Hazra, R., Wüthrich, M.: Inhomogeneous long-range percolation for real-life network modeling. Risks **3**, 1–23 (2015)
6. Deprez, P., Wüthrich, M.V.: Scale-free percolation in continuum space. Commun. Math. Stat. **7**(3), 269–308 (2019)
7. Gandolfi, A., Keane, M.S., Newman, C.M.: Uniqueness of the infinite component in a random graph with applications to percolation and spin glasses. Probab. Theory Rel. Fields **92**(4), 511–527 (1992). https://doi.org/10.1007/BF01274266
8. Gracar, P., Grauer, A., Lüchtrath, L., Mörters, P.: The age-dependent random connection model. Queueing Syst. **93**(3–4), 309–331 (2019)

9. Gracar, P., Grauer, A., Mörters, P.: Ultrasmallness in spatial random graphs. In preparation (2020)
10. Gracar, P., Heydenreich, M., Mönch, C., Mörters, P.: Recurrence versus transience for weight-dependent random connection models. arXiv e-prints, November 2019. arXiv:1911.04350
11. Gracar, P., Lüchtrath, L., Mörters, P.: Percolation phase transition in the weight-dependent random connection model. arXiv e-prints, March 2020. arXiv:2003.04040
12. HALL, P.: On continuum percolation. Ann. Probab. **13**, 1250–1266 (1985)
13. Heydenreich, M., Hulshof, T., Jorritsma, J.: Structures in supercritical scale-free percolation. Ann. Appl. Probab. **27**(4), 2569–2604 (2017)
14. Heydenreich, M., van der Hofstad, R., Last, G., Matzke, K.: Lace expansion and mean-field behavior for the random connection model. Preprint arXiv: 1908.11356 [math.PR] (2019)
15. Hirsch, C.: From heavy-tailed Boolean models to scale-free Gilbert graphs. Braz. J. Probab. Stat. **31**(1), 111–143 (2017)
16. Jacob, E., Mörters, P.: Spatial preferential attachment networks: power laws and clustering coefficients. Ann. Appl. Probab. **25**(2), 632–662 (2015)
17. Jacob, E., Mörters, P.: Robustness of scale-free spatial networks. Ann. Probab. **45**(3), 1680–1722 (2017)
18. Last, G., Penrose, M.D.: Lectures on the Poisson Process, vol. 7. Cambridge University Press, Cambridge (2018)
19. Last, G., Ziesche, S.: On the Ornstein-Zernike equation for stationary cluster processes and the random connection model. Adv. Appl. Probab. **49**(4), 1260–1287 (2017)
20. Meester, R., Penrose, M.D., Sarkar, A.: The random connection model in high dimensions. Stat. Probab. Lett. **35**(2), 145–153 (1997)
21. Meester, R., Roy, R.: Uniqueness of unbounded occupied and vacant components in Boolean models. Ann. Appl. Probab. **4**(3), 933–951 (1994)
22. Meester, R., Roy, R.: Continuum Percolation. Cambridge University Press, Cambridge (1996)
23. Michelitsch, T., Collet, B., Riascos, A.P., Nowakowski, A., Nicolleau, F.: On recurrence and transience of fractional RandomWalks in lattices. In: Altenbach, H., Pouget, J., Rousseau, M., Collet, B., Michelitsch, T. (eds.) Generalized Models and Non-classical Approaches in Complex Materials 1. ASM, vol. 89, pp. 555–580. Springer, Cham (2018). https://doi.org/10.1007/978-3-319-72440-9_29
24. Penrose, M.D.: Random Geometric Graphs. Oxford University Press, Oxford (2003)
25. Penrose, M.D.: Connectivity of soft random geometric graphs. Ann. Appl. Probab. **26**(2), 986–1028 (2016)
26. Peres, Y.: Probability on trees: an introductory climb. In: Bernard, P. (ed.) Lectures on Probability Theory and Statistics. LNM, vol. 1717, pp. 193–280. Springer, Heidelberg (1999). https://doi.org/10.1007/978-3-540-48115-7_3
27. Viswanathan, G., Raposo, E., da Luz, M.: Lévy flights and superdiffusion in the context of biological encounters and random searches. Phys. Life. Rev. **5**(3), 133–150 (2008)
28. Yukich, J.E.: Ultra-small scale-free geometric networks. J. Appl. Probab. **43**(3), 665–677 (2006)

The Iterated Local Directed Transitivity Model for Social Networks

Anthony Bonato[1](✉), Daniel W. Cranston[2], Melissa A. Huggan[1],
Trent Marbach[1], and Raja Mutharasan[1]

[1] Ryerson University, Toronto, ON, Canada
{abonato,melissa.huggan,trent.marbach,rmutharasan}@ryerson.ca
[2] Virginia Commonwealth University, Richmond, VA, USA
dcranston@vcu.edu

Abstract. We introduce a new, deterministic directed graph model for social networks, based on the transitivity of triads. In the Iterated Local Directed Transitivity (ILDT) model, new nodes are born over discrete time-steps and inherit the link structure of their parent nodes. The ILDT model may be viewed as a directed graph analog of the ILT model for undirected graphs introduced in [4]. We investigate network science and graph-theoretical properties of ILDT digraphs. We prove that the ILDT model exhibits a densification power law, so that the digraphs generated by the models densify over time. The number of directed triads are investigated, and counts are given of the number of directed 3-cycles and transitive 3-cycles. A higher number of transitive 3-cycles are generated by the ILDT model, as found in real-world, on-line social networks that have orientations on their edges. We discuss the eigenvalues of the adjacency matrices of ILDT digraphs. We finish by showing that in many instances of the chosen initial digraph, the model eventually generates digraphs with Hamiltonian directed cycles.

1 Introduction

Real-world, complex networks contain numerous mechanisms governing link formation. *Balance theory* (or *structural balance theory*) in social network analysis cites several mechanisms to complete triads (that is, subgraphs consisting of three nodes) in social networks [9,11]. A central mechanism in balance theory is *transitivity*: if x is a friend of y, and y is a friend of z, then x is a friend of z; see, for example, [16]. Directed networks of ratings or trust scores and models for their propagation were first considered in [10]. *Status theory* for directed networks, first introduced in [14], was motivated by both trust propagation and balance theory. While balance theory focuses on likes and dislikes, status theory posits that a directed link indicates that the creator of the link views the recipient as having higher status. For example, on Twitter or other social media, a directed

The first author acknowledges funding from an NSERC Discovery grant, while the third author acknowledges support from an NSERC Postdoctoral Fellowship.

© Springer Nature Switzerland AG 2020
B. Kamiński et al. (Eds.): WAW 2020, LNCS 12091, pp. 111–123, 2020.
https://doi.org/10.1007/978-3-030-48478-1_8

link captures one user following another, and the person they follow may be of higher social status. Evidence for status theory was found in directed networks derived from Epinions, Slashdot, and Wikipedia [14]. For other applications of status theory and directed triads in social networks, see also [12, 18].

The *Iterated Local Transitivity* (*ILT*) model introduced in [3, 4] and further studied in [2, 17], simulates social networks and other complex networks. Transitivity gives rise to the notion of *cloning*, where a node x is adjacent to all of the neighbors of y. Note that in the ILT model, the nodes have local influence within their neighbor sets. Although the model graph evolves over time, there is still a memory of the initial graph hidden in the structure. The ILT model simulates many properties of social networks. For example, as shown in [4], graphs generated by the model densify over time and exhibit bad spectral expansion. In addition, the ILT model generates graphs with the small-world property, which requires the graphs to have low diameter and dense neighbor sets.

In the present work, we introduce a directed analog of the ILT model, where nodes are added and copy the in- and out-neighbors of existing nodes. The model simulates link creation in social networks, where new actors enter the network, and directed edges are added via transitivity through the lens of status theory. For example, in link formation in a directed social network such as Twitter, a new user may reciprocally follow an existing one, then in turn follow their followers. We consider the simplified setting where new nodes copy all of the links of their parent node. Our model, and iterated models more generally [2], provide counterparts for earlier studied random models for complex networks involving copying [13] or duplication [8].

More formally, the *Iterated Local Directed Transitivity* (*ILDT*) model is deterministically defined over discrete time-steps as follows. The only parameter of this deterministic model is the initial digraph $G = G_0$. For a non-negative integer t, the graph G_t represents the digraph at time-step t. Suppose that the directed graph G_t has been defined for a fixed time $t \geq 0$. To form G_{t+1}, for each $x \in V(G_t)$, add a new node x' called the *clone* of x. We refer to x as the *parent* of x', and x' as the *child* of x. Between x and x' we add a bidirectional arc, representing a reciprocal status (or friendship) relationship between them. For arcs (x, z) and (y, x) in G_t, we add arcs (x', z) and (y, x'), respectively, in G_{t+1}. See Fig. 1. We refer to G_t as an *ILDT digraph*. Note that the clones form an independent set in G_{t+1}.

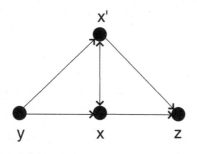

Fig. 1. Adding a clone x' in ILDT.

In Fig. 2, we illustrate several time-steps of the ILDT model beginning with the directed 3-cycle.

The paper is organized as follows. In Sect. 2, we prove that the ILDT model exhibits a densification power law, so that the digraphs generated by the model densify over time. The number of directed triads are investigated in Theorem 2, and precise counts are given of the number of directed 3-cycles and transitive cycles. These counts are contrasted, and it is shown that the transitive cycles are more abundant (as is the case with social networks; see [14]). We include a discussion of the eigenvalues of the adjacency matrices of ILDT directed graphs. In Sect. 3, the eigenvalues of ILDT directed graphs are investigated. In Sect. 4, we explore directed cycles of larger order in ILDT graphs. We show that ILDT digraphs are acyclic if the initial digraph is such, and that for many instances of initial digraphs, the model eventually generates graphs with Hamiltonian directed cycles. We finish with open problems.

For a general reference on graph theory, the reader is directed to [19]. For background on social and complex networks, see [1,5,7]. Throughout the paper, we consider finite, directed graphs with bidirectional edges allowed. We refer to

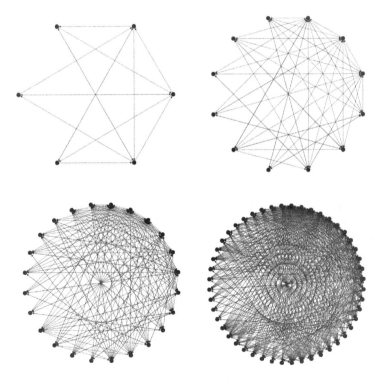

Fig. 2. The ILDT graphs G_t, with $t = 1, 2, 3, 4$ of the ILDT model, where the initial graph is the directed 3-cycle.

a directed edge as an *arc*. For nodes x and y of a graph, if there is an arc between x and y, we denote it by (x, y); if there is a *bidirectional arc* between x and y we denote such an arc with usual graph theoretic notation of xy. When counting the number of arcs within a graph, bidirectional arcs each contribute 2 to the final count. We use $\log n$ to be the logarithm of n in base 2.

2 Densification and Triad Counts

As we referenced in the introduction, social networks *densify*, in the sense that the ratio of their number of arcs to nodes tends to infinity over time [15]. We show in this section that the ILDT model always generates digraphs that densify, and we give a precise statement below of its densification power law. As a phenomenon specific to digraphs, we consider the differing counts of directed and transitive 3-cycles in graphs generated by the model.

The number of nodes of G_t is denoted n_t, the number of arcs is denoted e_t, and the number of bidirectional arcs is denoted b_t. Note that e_t contains two arcs for each bidirectional arc. We establish elementary but important recursive formulas for these parameters.

Lemma 1. *Let G_0 be a digraph with n_0 nodes, e_0 arcs, and b_0 bidirectional arcs. For all $t \geq 1$, we have the following:*

1. *$n_t = 2^t n_0$;*
2. *$e_t = 3e_{t-1} + 2n_{t-1}$; and*
3. *$b_t = 3b_{t-1} + n_{t-1}$.*

Proof. Item (1) follows immediately as the number of nodes doubles in each time-step. For item (2), for each node in G_{t-1} with $t > 0$, after cloning there will be a bidirectional arc between each parent and their child. These count as $2n_{t-1}$-many arcs. For every arc (x, y) in G_{t-1}, arcs (x, y') and (x', y) appear in G_t. Hence, for every arc in G_{t-1}, three arcs of G_t are generated. Summing these two counts gives the desired expression for e_t. Item (3) follows analogously to (2), except that we count the bidirectional edges once; hence, there are n_{t-1}-many bidirectional arcs.   □

We now state the densification power law for ILDT graphs. For positive integer-valued functions f_t and g_t, we use the expression $f_t \sim g_t$ if f_t/g_t tends to 1 as t tends to ∞.

Corollary 1. *In the ILDT model, we have that*

$$\frac{e_t}{n_t} \sim \left(\frac{3}{2}\right)^t \frac{(e_0 + 2n_0)}{n_0}.$$

In particular, we have that $e_t \sim C \cdot (n_t)^a$, where $a = \log 3$ and $C = \frac{e_0 + 2n_0}{(n_0)^a}$.

Proof. By Lemma 1, we have that

$$e_t = 3^t e_0 + 3^{t-1} 2^1 n_0 + 3^{t-2} 2^2 n_0 + \ldots + 3^1 2^{t-1} n_0 + 2^t n_0$$

$$= 3^t e_0 + 3^{t-1} 2 n_0 \left(\frac{1 - \left(\frac{2}{3}\right)^t}{1 - \frac{2}{3}} \right)$$

$$= 3^t (e_0 + 2n_0) - 2^t (2n_0).$$

We then derive that

$$\frac{e_t}{n_t} = \frac{3^t (e_0 + 2 \cdot n_0) - 2^t (2n_0)}{2^t n_0} \sim \left(\frac{3}{2}\right)^t \frac{(e_0 + 2n_0)}{n_0},$$

and the result follows.   □

We next consider 3-node subgraph counts in the ILDT model, where we find a higher number of transitive 3-cycles relative to directed 3-cycles. A similar phenomenon was found in directed network samples in social media such as Epinions, Slashdot, and Wikipedia, where transitive 3-cycles appear much more commonly than directed 3-cycles; see [14]. If the initial graph has no directed 3-cycles, then our results show there are none at any time-step in the evolution of the model. For nodes x, y, z, if there exists a cycle with arcs (x, y), (y, z), and (z, x), then we abbreviate this directed 3-cycle to (x, y, z). We take the directed 3-cycles (x, y, z) and (y, z, x) to be the same cycle. For a 3-cycle where the arcs (x, y), (y, z), and (x, z) are present, we denote this transitive 3-cycle by xyz. A *bidirectional 3-cycle* is one that consists of three bidirectional arcs. Although, strictly speaking, a bidirectional 3-cycle contains six transitive 3-cycles and two directed 3-cycles, we will distinguish these by asserting that each transitive and directed 3-cycle must contain at least one non-bidirectional arc.

Theorem 1. *In the graph G_t, let D_t be the number of directed 3-cycles, T_t be the number of transitive 3-cycles, and B_t be the number of bidirectional 3-cycles. We then have that*

1. $D_{t+1} = 4D_t$;
2. $T_{t+1} = 4T_t + 4(e_t - 2b_t)$; *and*
3. $B_{t+1} = 4B_t + 2b_t$.

Proof. For item (1), consider a directed 3-cycle in a graph G_t, labeled (a, b, c). Each node can be replaced by its clone to produce a new directed 3-cycle; hence, (a', b, c), (a, b', c), and (a, b, c') are all directed 3-cycles generated by the directed 3-cycle (a, b, c) from G_t. Thus, for each 3-cycle in G_t, there are four directed 3-cycles in G_{t+1}. Therefore, $D_{t+1} \geq 4D_t$.

To establish the upper bound, suppose, by way of contradiction, that there exists another directed 3-cycle in G_{t+1} which was not previously counted. Such a directed 3-cycle cannot involve only parent nodes (since it would be from G_t) and it also cannot involve two clones because clones form an independent set. Hence, it must involve two nodes from G_t and one of the clones, call it (a, d', c). But since d' has all the same adjacencies as d, this implies that (a, d, c) is a directed 3-cycle which is in G_t, a contradiction, or that a, d, c are not all distinct. In the later case, we can assume that $a = d$, and so the directed 3-cycle includes arcs (a, c) and (c, a'), implying that 3-cycle is bidirectional, which is not counted in D_{t+1}. Hence, $D_{t+1} = 4D_t$.

To prove item (2), observe that for every non-bidirectional arc (x, y) in G_t, there will be four transitive 3-cycles in G_{t+1} formed with their clones using bidirectional arcs: $xx'y$, $x'xy$, xyy', and $xy'y$. Thus, we add $4(e_t - 2b_t)$ to the count of T_{t+1} (note that $e_t - 2b_t$ counts the number of non-bidirectional arcs).

Existing transitive 3-cycles from G_t also exist in G_{t+1}, and substituting a clone for its parent will produce a new transitive 3-cycle. Thus, each original transitive 3-cycle in G_t gives four transitive 3-cycles in G_{t+1} (namely, the cycle itself, and three produced from clone substitution). It is straightforward to check that these are the only transitive 3-cycles in G_{t+1}. Therefore, $T_t + 3T_t$ contributes to the total count of transitive 3-cycles. We then have that $T_{t+1} = 4T_t + 4(e_t - 2b_t)$.

Item (3) can be shown similarly, as each bidirectional 3-cycle of G_t will correspond to four bidirectional 3-cycles in G_{t+1}. Each bidirectional arc xy in G_t will correspond to two unique bidirectional 3-cycles in G_{t+1} formed from the nodes x, x', y, y'. □

We now present an exact expressions for D_t, T_t, and B_t.

Theorem 2. *In the ILDT digraph G_t, we have that*

1. $D_t = 4^t D_0$;
2. $T_t = 4^t T_0 + 4(4^t - 3^t)(e_0 - 2b_0)$; and
3. $B_t = 4^t B_0 + 2b_0(4^t - 3^t) + n_0(4^t - 2 \cdot 3^t + 2^t)$.

Proof. Item (1) follows from Theorem 1 (1) by induction. For (2), by Theorem 1 (2), we derive that

$$T_t = 4^t T_0 + \sum_{i=1}^{t} 4^i(e_{t-i} - 2b_{t-i}).$$

By the proof of Corollary 1, along with a similar argument for b_t, we have that

$$e_{t-i} = 3^{t-i}(e_0 + 2n_0) - 2^{t-i+1}n_0,$$
$$b_{t-i} = 3^{t-i}(b_0 + n_0) - 2^{t-i}n_0, \text{ and so} \tag{1}$$
$$e_{t-i} - 2b_{t-i} = 3^{t-i}(e_0 - 2b_0). \tag{2}$$

From (2), we derive that

$$T_t = 4^t T_0 + \sum_{i=1}^{t} 4^i(3^{t-i}(e_0 - 2b_0))$$

and item (2) follows by summing the geometric series.

For item (3), we have that $B_t = 4^t B_0 + \frac{1}{2} \sum_{i=1}^{t} 4^i b_{t-i}$. From Eq. (1), we find (in a way analogous to T_t) the desired expression for B_t.   □

We consider next the ratio of D_t and T_t, which gives more explicit estimates on the relative abundance of transitive versus directed 3-cycles. By Theorem 2, we have that

$$\frac{D_t}{T_t} = \frac{4^t D_0}{4^t T_0 + 4(4^t - 3^t)(e_0 - 2b_0)} \sim \frac{D_0}{T_0 + 4(e_0 - 2b_0)}.$$

Hence, the ratio $\frac{D_t}{T_t}$ may be made as small as we like by choosing an appropriate initial digraph.

3   Eigenvalues

We next consider eigenvalues of the adjacency matrices of ILDT digraphs. Spectral graph theory is a well-developed area for undirected graphs (see [6]) but less so for directed graphs (where the eigenvalues may be complex numbers with non-zero imaginary parts). We observe that if G_t has adjacency matrix A, then G_{t+1} has the following adjacency matrix:

$$\begin{pmatrix} A & A+I \\ A+I & 0 \end{pmatrix},$$

where I and 0 are the appropriately sized identity and zero matrices, respectively. The following recursive formula (analogous to the one in the ILT model) allows us to determine all the eigenvalues of ILDT graphs from the spectrum of the initial graph. We have the following theorem from [4].

Theorem 3. *Let $t \geq 0$. If ρ is an eigenvalue of the adjacency matrix of G_t, then the eigenvalues of the adjacency matrix of G_{t+1} are*

$$\frac{\rho \pm \sqrt{\rho^2 + 4(\rho + 1)^2}}{2}.$$

It is of interest to consider properties of the distribution of eigenvalues arising from ILDT digraphs graphs in the complex plane. For this, we consider the special case of G_0 a directed 3-cycle, which has eigenvalues the 3rd roots of unity. The resulting eigenvalue distribution of these ILDT digraphs suggests a rich structure. We plot the eigenvalues corresponding to G_t for $1 \leq t \leq 5$ in Fig. 3a.

If ρ is an eigenvalue of the adjacency matrix of G_t and ρ is large in magnitude, then there is an eigenvalue of the adjacency matrix of G_{t+1} that is approximately equal to $((1+\sqrt{5})/2)\rho$, and in a similar way there is an eigenvalue of the adjacency matrix of $G_{t+\alpha}$ that is approximately equal to $((1 + \sqrt{5})/2)^\alpha \rho$. We *normalize* the eigenvalues corresponding to G_t by dividing them by $((1+\sqrt{5})/2)^t$ for $1 \leq t \leq 5$ in Fig. 3b.

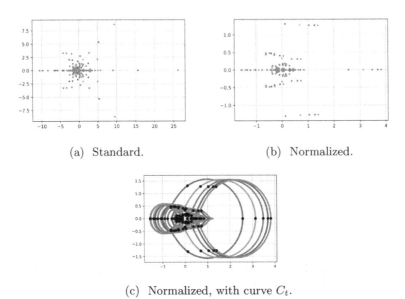

(a) Standard.   (b) Normalized.

(c) Normalized, with curve C_t.

Fig. 3. Eigenvalues in the complex plane of ILDT digraphs G_t, where $1 \leq t \leq 5$. Colors distinguish the time-steps. Figures (a) depicts the eigenvalues, (b) the normalized eigenvalues, and (c) depicts the curves C_t.

Let C_0 be the circle in the complex plane of radius 1 centered at the origin. By applying the function $f(z) = (z \pm \sqrt{z^2 + 4(z+1)^2})/2$ iteratively t times to the points of C_0, we obtain a curve C_t in the complex plane. If we let G_0 be the directed n-cycle, then the eigenvalues of G_t lie on C_t. In Fig. 3c, we include C_t and the eigenvalues of G_t after normalization by dividing them by $((1+\sqrt{5})/2)^t$, where G_0 is the directed 3-cycle. The curve C_t was plotted after normalization for $t \leq 30$, and there was no noticeable difference between the time-steps $t = 15$ and $t = 30$. The structure after 30 iterations is provided in Fig. 4. We suspect that as t approaches infinity, the normalization of C_t approaches a specific, limiting curve. As a result, the normalized mapping applied to the nth roots of unity would approach limiting points, and these limiting points can be calculated from the curve.

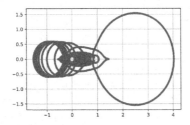

Fig. 4. The mapping $f(z)$ applied to the complex unit circle over 30 iterations, after normalization.

4   Directed Cycles

As a consequence of Theorem 2, there are no directed 3-cycles in an ILDT digraph unless there is one present in the initial graph. We generalize this property in the following result. Note that our directed cycles in the theorem are oriented, and so do not include directed 2-cycles. However, we are allowed to include bidirected edges (traversed in a single-direction) as part of our directed cycles.

Theorem 4. *For all $t \geq 0$, the digraph G_t contains an oriented directed cycle if and only if G_{t+1} contains an oriented directed cycle.*

Proof. The forward implication is immediate, so we focus on the reverse implication. Let $(a_1, \ldots, a_k)$ be a directed cycle of length k in G_{t+1}. Define a function $f : V(G_{t+1}) \to V(G_t)$ that maps a clone in G_{t+1} to its parent node, and acts as the identity mapping, otherwise. If (a_i, a_{i+1}) is an arc in G_{t+1}, then $(f(a_i), f(a_{i+1}))$ is an arc in G_t. The subgraph induced by the edges of the closed directed walk $(f(a_1), \ldots, f(a_k))$ has in-degree equal to out-degree at every node (counting the multiplicity of the edges in the walk) precisely because it is a closed directed walk. Each time we visit a node on the walk we contribute one to the in-degree and one to the out-degree. Hence, $(f(a_1), \ldots, f(a_k))$ decomposes into an edge-disjoint collection of directed cycles (none of which is a directed 2-cycle, by hypothesis). Hence, G_t contains a cycle. □

We turn next to directed Hamiltonian cycles; that is, directed cycles visiting each node exactly once. Note that while we do not expect directed Hamiltonian cycles in real-world social networks, the emergence of this property in ILDT graphs is of graph-theoretical interest in its own right. For this, we first prove the following theorem on Hamiltonian paths in ILT undirected graphs. For a graph G, we use the notation $\text{ILT}_t(G)$ for the ILT graph resulting at time t if $G_0 = G$; analogous notation is used for ILDT graphs. We use the notation $G[S]$ for the subgraph induced by nodes S in G.

In the following lemma with G_0 chosen as K_1, we label the node of G_0 as 0, and its unique child in G_1 as 1.

Lemma 2. *Fix $t \geq 1$ and let $G_t = ILT_t(K_1)$. For every clone $v \in V(G_t)$, there is a Hamiltonian path in G_t from v to 0.*

Proof. We use induction on $t \geq 1$. The base case $t = 1$ is straightforward, since $G_1 \cong K_2$. For the induction step, we assume $t \geq 2$. We label the node of G_0 as 0, and its unique child in G_1 as 1. Note that we can partition $V(G_t)$ into V_0 and V_1 such that $0 \in V_0$, $1 \in V_1$ and $G_t[V_0] \cong G_t[V_1] \cong G_{t-1}$. To see this, consider the ILDT process as starting with each of the nodes 0 and 1 independently. For $i = 1, 2$, the set V_i consists of all clones over subsequent time-steps starting with the initial vertex i.

First suppose that $v \in V_1$, and choose an arbitrary $w \in V_0$. By the induction hypothesis, there exists a Hamiltonian path P_0 in $G_t[V_0]$ with endpoints w and 0. Similarly, there exists a Hamiltonian path P_1 in $G_t[V_1]$ with endpoints v and

1. Let $P = P_1 + 1w + P_0$; now P is the desired Hamiltonian path in G_t from v to 0.

Suppose instead that $v \in V_0$, and choose an arbitrary $w \in V_1$. By the induction hypothesis, there exists a Hamiltonian path P_0 in $G_t[V_0]$ with endpoints v and 0. Similarly, there exists a Hamiltonian path P_1 in $G_t[V_1]$ with endpoints w and 1. Let x be the neighbor of 0 on P_0. Let $P = P_0 - x0 + x1 + P_1 + w0$. Now P is the desired Hamiltonian path in G_t from v to 0 (Fig. 5).   □

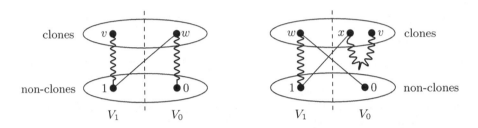

Fig. 5. The two cases in the proof of Lemma 2, wavy lines represent paths.

In a digraph D, a closed spanning walk $\mathcal{C}$ (that respects the orientations of D) is *nice* if for each edge $v_i v_j \in E(\mathcal{C})$ either (i) $v_i v_j$ is the last edge departing v_i on $\mathcal{C}$ or (ii) $v_i v_j$ is the first edge entering v_j on $\mathcal{C}$; possibly both (i) and (ii) hold for some edges of $\mathcal{C}$. The *max frequency*, written $s(\mathcal{C})$, of a nice walk $\mathcal{C}$ is the largest number of times that any node appears in $\mathcal{C}$.

Theorem 5. *If D is a digraph with a nice walk $\mathcal{C}$ and $t \in \mathbb{Z}^+$ such that $2^{t-1} \geq s(\mathcal{C})$, then D_t has a directed Hamiltonian cycle.*

Proof. Let $D_t = \mathrm{ILDT}_t(D)$. We construct a Hamiltonian cycle in D_t by the algorithm below. We assume that the nodes of D are $\{v_1, \ldots, v_n\}$ (each v_i may appear many times on $\mathcal{C}$) and that the nodes of D_t are partitioned into $V_1, \ldots, V_n$ (where V_i consists of v_i and all its *descendants*; that is, nodes that resulted by iterated cloning of v_i). Intuitively, we use $\mathcal{C}$ to ensure that our walk $\mathcal{H}$ visits each V_i at least once and we use each P_i to ensure that we visit all remaining vertices of V_i the last time that $\mathcal{C}$ visits v_i, in D.

Initialization: Pick an arbitrary node v_i on $\mathcal{C}$. Choose a clone $w \in V_i$. Start $\mathcal{H}$ at w. Let P_i be a Hamiltonian path in V_i that starts at w and ends at 0 (in V_i), by Lemma 2 (with vertex 0 defined as in Lemma 2). Always v_i, v_j refers to vertices of D and vertices of D_t are denoted by w, x, or 0.

Iteration: Assume that $\mathcal{H}$ currently ends at some clone $w \in V_i$ (for some i) and path P_i is defined, possibly from the initialization. If we have followed all edges of $\mathcal{C}$, then halt and output $\mathcal{H}$. Otherwise, let $e = v_i v_j$ denote the next edge of $\mathcal{C}$. (1) If e is the last edge leaving v_i on $\mathcal{C}$, then follow P_i from w to 0 in V_i; otherwise, move to the next node on P_i. (2) If P_j is undefined (we have not yet visited v_j on $\mathcal{C}$), then follow an edge to an arbitrary clone x in V_j. In this case, define P_j to be a Hamiltonian path in V_j with endpoints x and 0; such a P_j exists by Lemma 2. If P_j is defined, then follow an edge from the current node of $\mathcal{H}$ to the next node on P_j (in V_j). (When e is the final edge of $\mathcal{C}$, we return to the node of V_j where we started, which finishes $\mathcal{H}$.)

This completes the algorithm for constructing $\mathcal{H}$ from $\mathcal{C}$. We prove its correctness in two steps. First, we show that if the algorithm completes, then it constructs the desired Hamiltonian cycle $\mathcal{H}$. Second, we show that the algorithm does indeed complete. Suppose the algorithm completes successfully. Since $\mathcal{C}$ is a spanning walk, it visits every node $v_i \in V(D)$. Thus, $\mathcal{H}$ visits every V_i. The final time that $\mathcal{H}$ visits a V_i it visits every remaining node on P_i. Thus, $\mathcal{H}$ visits every node in $\bigcup V_i = V(D_t)$; that is, $\mathcal{H}$ is spanning in D_t.

Now we must show that the algorithm completes successfully. Every time $\mathcal{H}$ leaves a V_i it does so from a non-clone, and every time $\mathcal{H}$ returns to a V_j it returns to a clone (that has not been visited before). The number of clones in each V_j is 2^{t-1}, so this is possible precisely because $s(\mathcal{C}) \geq 2^{t-1}$. Now we need to check that D_t has the necessary edges between V_i and V_j. Since $\mathcal{C}$ is nice, each edge $e = v_i v_j \in E(\mathcal{C})$ satisfies that either (i) $v_i v_j$ is the last edge leaving v_i on $\mathcal{C}$ or (ii) $v_i v_j$ is the first edge entering v_j on $\mathcal{C}$. In (i), $\mathcal{H}$ leaves V_i from node 0, which has edges to every node of V_j. In (ii), any edge from a non-clone of V_i to a clone of V_j suffices, since we will define P_j as starting from this clone of V_j (and since $\mathcal{C}$ has never before visited v_j). $\qquad\square$

The following result on the Hamiltonicity of the ILT model was first proven in [2], and we give an alternative proof as a corollary of Theorem 5.

Corollary 2. *If G is a connected undirected graph and $t = \log|V(G)|$, then $ILT_t(G)$ is Hamiltonian.*

Proof. We form a digraph D from G by replacing each undirected edge vw with arcs (v, w) and (w, v). We construct a nice spanning closed walk of D and apply Theorem 5. Choose an arbitrary node $v \in D$ and form $\mathcal{C}$ by recording each edge followed in a depth-first traversal of D (including to what we call back-tracking edges). Consider an edge $vw \in E(\mathcal{C})$. If w has never been visited before, then vw satisfies (ii) in the definition of nice. If w has been visited before, then it is straightforward to check that vw satisfies (i) in the definition (precisely because $\mathcal{C}$ arose from a depth-first traversal of D). $\qquad\square$

We conjecture that for every strongly connected digraph D there exists an integer t such that $ILT_t(D)$ has a Hamiltonian cycle. In a sense, this conjecture is best possible, since if $ILT_t(D)$ is Hamiltonian for some t, then D must be strongly connected. Namely, if there exist $v_i, v_j \in V(D)$ such that D has no directed path from v_i to v_j, then $ILT_t(D)$ has no directed path from V_i to V_j, so $ILT_t(D)$ is not Hamiltonian. We suspect that this conjecture can be proved by somehow modifying the proof of Theorem 5.

5   Conclusion and Further Directions

We introduced and analyzed the Iterated Local Directed Transitivity (ILDT) model for social networks, motivated by status theory, transitivity in triads, and the ILT model in the undirected case [4]. We proved that the ILDT model, as in social networks, generates graphs that densify over time. A count of the directed, transitive, and bidirectional 3-cycles was given, and it was shown that the 3-transitive cycles count may be far more abundant by choice of the initial graph of the model. We studied the eigenvalues of the adjacency matrices of ILDT graphs, with a discussion of the limiting distribution of eigenvalues of the directed 3-cycle. We concluded our results with an analysis of directed cycles in ILDT graphs and proved that in many instances of the initial graph, ILDT graphs have Hamiltonian cycles.

Given our limited space, we did not explore distance properties of the model, although we expect the model should generate small-world graphs, as is the case for ILT graphs. In the full version of the paper, it would be interesting to analyze the clustering coefficient, domination number, and degree distribution of ILDT graphs. The eigenvalues of ILDT graphs are worthy of further study, both in their limiting distribution in the complex plane and regarding their spectral expansion.

References

1. Bonato, A.: A Course on the Web Graph. Graduate Studies Series in Mathematics. American Mathematical Society, Providence (2008)
2. Bonato, A., Chuangpishit, H., English, S., Kay, B., Meger, E.: The iterated local model for social networks. Discrete Applied Mathematics (accepted)
3. Bonato, A., Hadi, N., Horn, P., Prałat, P., Wang, C.: A dynamic model for on-line social networks. In: Avrachenkov, K., Donato, D., Litvak, N. (eds.) WAW 2009. LNCS, vol. 5427, pp. 127–142. Springer, Heidelberg (2009). https://doi.org/10.1007/978-3-540-95995-3_11
4. Bonato, A., Hadi, N., Horn, P., Prałat, P., Wang, C.: Models of on-line social networks. Internet Math. **6**, 285–313 (2011)
5. Bonato, A., Tian, A.: Complex networks and social networks. In: Kranakis, E. (ed.) Social Networks. Mathematics in Industry, vol. 18, pp. 269–286. Springer, Heidelberg (2011). https://doi.org/10.1007/978-3-642-30904-5_12
6. Chung, F.R.K.: Spectral Graph Theory. American Mathematical Society, Providence (1997)
7. Chung, F.R.K., Lu, L.: Complex Graphs and Networks. American Mathematical Society, Providence (2006)
8. Chung, F.R.K., Dewey, G., Galas, D.J., Lu, L.: Duplication models for biological networks. J. Comput. Biol. **10**, 677–688 (2003)
9. Easley, D., Kleinberg, J.: Networks, Crowds, and Markets Reasoning About a Highly Connected World. Cambridge University Press, Cambridge (2010)
10. Guha, R.V., Kumar, R., Raghavan, P., Tomkins, A.: Propagation of trust and distrust. In: Proceedings of WWW (2004)
11. Heider, F.: The Psychology of Interpersonal Relations. John Wiley, New York (1958)

12. Huang, J., Shen, H., Hou, L., Cheng, X.: Signed graph attention networks. In: Proceedings of Artificial Neural Networks and Machine Learning – ICANN (2019)
13. Kumar, R., Raghavan, P., Rajagopalan, S., Sivakumar, D., Tomkins, A., Upfal, E.: Stochastic models for the web graph. In: Proceedings of the 41st IEEE Symposium on Foundations of Computer Science (2000)
14. Leskovec, J., Huttenlocher, D.P., Kleinberg, J.M.: Signed networks in social media. In: Proceedings of the ACM SIGCHI (2010)
15. Leskovec, J., Kleinberg, J., Faloutsos, C.: Graphs over time: densification laws, shrinking diameters and possible explanations. In: Proceedings of the 13th ACM SIGKDD International Conference on Knowledge Discovery and Data Mining (2005)
16. Scott, J.P.: Social Network Analysis: A Handbook. Sage Publications Ltd., London (2000)
17. Small, L., Mason, O.: Information diffusion on the iterated local transitivity model of online social networks. Discrete Appl. Math. **161**, 1338–1344 (2013)
18. Song, D., Meyer, D.A.: A model of consistent node types in signed directed social networks. In: Proceedings of the 2014 IEEE/ACM International Conference on Advances in Social Networks Analysis and Mining (2014)
19. West, D.B.: Introduction to Graph Theory, 2nd edn. Prentice Hall, Upper Saddle River (2001)

A Note on the Conductance of the Binomial Random Intersection Graph

Katarzyna Rybarczyk[1(✉)], Mindaugas Bloznelis[2], and Jerzy Jaworski[1]

[1] Adam Mickiewicz University in Poznań, Poznań, Poland
kryba@amu.edu.pl
[2] Vilnius University, Vilnius, Lithuania

Abstract. We establish the rapid mixing property of a binomial random intersection graph $\mathcal{G}(n, m, p)$ introduced in [11,17]. For this purpose we show that the conductance is bounded away from zero by a positive constant. We consider the range of parameters n, m, p where the edge density is just above the connectivity threshold (our graph is connected with high probability as $n, m \to +\infty$). We assume in addition that $np = \Theta(1)$.

Keywords: Random intersection graph · Mixing time · Conductance

1 Introduction

Model. Vertices $v_1, \ldots, v_n$ of an intersection graph represent subsets $\mathcal{W}(v_1), \ldots, \mathcal{W}(v_n)$ of an auxiliary set $\mathcal{W} = \{w_1, \ldots, w_m\}$. Two vertices v_i and v_j are called adjacent whenever the corresponding subsets intersect, $\mathcal{W}(v_i) \cap \mathcal{W}(v_j) \neq \emptyset$. In a random intersection graph the subsets $\mathcal{W}(v_1), \ldots, \mathcal{W}(v_n)$ are drawn at random. Random intersection graphs have attracted considerable attention in recent literature mainly as convenient models of real complex networks. For example, in the collaboration network two scientists are adjacent if they have co-authored a publication, in the consumer co-purchase network two customers are adjacent if they purchased similar products, etc. They also have important applications in networks' modelling, see e.g. [1,3–6,8,12,18].

In this note we focus on the binomial random intersection graph $\mathcal{G}(n, m, p)$ introduced in [11,17]. In the binomial random intersection graph the random sets $\mathcal{W}(v_1), \ldots, \mathcal{W}(v_n)$ are independent and identically distributed. Moreover, each $w \in \mathcal{W}$ is included in $\mathcal{W}(v_i)$ independently at random with probability p for every v_i, $1 \leq i \leq n$. By $\mathcal{V} = \{v_1, \ldots, v_n\}$ we denote the vertex set. Elements of the auxiliary set $\mathcal{W}$ are called attributes. We will consider a sequence of random intersection graphs $\mathcal{G}_n = \mathcal{G}(n, m, p)$, where $m = m_n \to +\infty$ and $p = p_n \to 0$ as $n \to +\infty$. We assume that $np = \Theta(1)$ and $mp(1 - e^{-np}) - \ln n \to +\infty$. The latter relation implies that the event that $\mathcal{G}(n, m, p)$ is connected has probability $1 - o(1)$, see [15,17].

K. Rybarczyk—Supported by NCN (National Science Center) grant 2014/13/D/ST1/01175.

B. Kamiński et al. (Eds.): WAW 2020, LNCS 12091, pp. 124–134, 2020.
https://doi.org/10.1007/978-3-030-48478-1_9

Motivation. The conductance of a graph is related to problems concerning random walks on graphs [16]. In the accompanying paper [2] we use, among others, this result to establish a first order asymptotic to the cover time of $\mathcal{G}(n, m, p)$ in the range of parameters considered in Theorem 1. We believe that understanding the random walk performance in random intersection graphs is an interesting and important question as random intersection graphs are perceived as convenient models of real complex networks [1,18]. The binomial random intersection graph $\mathcal{G}(n, m, p)$ is a particular representative of the family of random intersection graphs and results concerning this model are the first step towards understanding the random walks in other random intersection graph models.

Result. The conductance $\Phi(G)$ of a connected graph $\mathcal{G}$ with vertex set $\mathcal{V}$ is defined as follows

$$\Phi(G) = \min_{\mathcal{S} \subset \mathcal{V}, |\mathcal{S}| \leq |\mathcal{V}|/2} \frac{e(\mathcal{S}, \bar{\mathcal{S}})}{2e(\mathcal{S}, \mathcal{S}) + e(\mathcal{S}, \bar{\mathcal{S}})},$$

where $e(\mathcal{S}, \mathcal{S})$ is the number of edges induced by set $\mathcal{S} \subset \mathcal{V}$ and $e(\mathcal{S}, \bar{\mathcal{S}})$ is the number of edges between $\mathcal{S}$ and $\bar{\mathcal{S}} := \mathcal{V} \setminus \mathcal{S}$. The conductance depicts how "narrow" are "bottlenecks" which delay the mixing of a random walk on a graph and is closely related to the mixing time of a random walk, see [16].

In Theorem 1 below we show that $\Phi(\mathcal{G}(n, m, p))$ is bounded away from zero by a positive constant as $n \to +\infty$. This implies that the mixing time of the simple random walk on $\mathcal{G}(n, m, p)$ is $O(\ln n)$, see [16]. In particular, the random intersection graph $\mathcal{G}(n, m, p)$ has the rapid mixing property.

Before formulating our result we mention that all limits in the paper are taken as $n \to \infty$. Throughout the paper we use standard asymptotic notation $o(\cdot)$, $O(\cdot)$, $\Omega(\cdot)$, $\Theta(\cdot)$, and $\asymp$ defined as in [10].

Theorem 1. *Let $c_2 > c_1 > 0$ and $c_3 > 0$. Let $n \to +\infty$. Let $m = m(n) \to +\infty$, $p = p(n) \to 0$ and $c = c(n) > 1$ be such that $c_1 \leq np \leq c_2$ and $c \leq c_3$, and*

$$mp(1 - (1 - p)^{n-1}) = c \ln n, \quad \text{and} \quad (c - 1) \ln n \to \infty. \tag{1}$$

Then

$$\Pr\{\Phi(\mathcal{G}(n, m, p)) > 0.05\} = 1 - o(1). \tag{2}$$

Related Work. In earlier papers [13,14] the $O(\ln n)$ upper bound on the mixing time of $\mathcal{G}(n, m, p)$ has been shown for the model's parameters $p = 4m^{-1} \ln n$ and $m = n^\alpha$, where $\alpha \leq 1$ is fixed. Note that for $m = n^\alpha$, $\alpha \leq 1$, the connectivity threshold is at $p = m^{-1} \ln n$. Interestingly, [13] lowerbounds the conductance, but of the related bipartite graph (see $\mathcal{B}(n, m, p)$ in Sect. 2 below) instead of $\mathcal{G}(n, m, p)$. In Theorem 1 we concentrate on the case where $m \asymp n \ln n$ therefore the range of parameters n, m, p considered here does not intersect with that of [13,14]. We note that the random intersection graph considered in Theorem 1 is a union of m randomly located cliques (for each $w \in \mathcal{W}$ the set $\{v: w \in \mathcal{W}(v)\}$ induces a clique in $\mathcal{G}(n, m, p)$). The sizes of these cliques are independent binomial random variables with the common distribution $\mathrm{Bin}(n, p)$.

Further Work. The result concerning conductance of the binomial random intersection graph $\mathcal{G}(n, m, p)$ is a first step towards a better understanding of random walks on random intersection graphs. It is essential in finding the expected cover time of a random walk on $\mathcal{G}(n, m, p)$ [2]. It is interesting and technically challenging problem to determine the cover time of (giant components of) more general models of random intersection graphs including those with non-vanishing clustering coefficient and power law degree distribution [1, 18].

2   Notation and Idea of the Proof

By $\mathrm{Bin}(n, p)$ we denote the binomial distribution with parameters n, p. We also use the phrase "with high probability" to say that the probability of a considered event tends to one as n tends to infinity. All inequalities hold for n large enough. If it does not influence the result, we consequently omit $\lfloor \cdot \rfloor$ and $\lceil \cdot \rceil$ for the sake of clarity of presentation. By $A_1, A_2, \ldots$ we denote positive constants that do not depend on n.

Let $\mathcal{B}(n, m, p)$ be the random bipartite graph with the bipartition $(\mathcal{V}, \mathcal{W})$, where each $v \in \mathcal{V}$ and $w \in \mathcal{W}$ are linked independently with probability p. Random intersection graph $\mathcal{G} = \mathcal{G}(n, m, p)$ with the vertex set $\mathcal{V}$ is obtained from $\mathcal{B} = \mathcal{B}(n, m, p)$ as follows: $v, v' \in \mathcal{V}$ are adjacent in $\mathcal{G}$ whenever v, v' have a common neighbour in $\mathcal{B}$. Now $\mathcal{W}(v)$ and $\mathcal{V}(w)$ stand for the sets of neighbours in $\mathcal{B}$ of vertices $v \in \mathcal{V}$ and $w \in \mathcal{W}$ respectively. Clearly, $\mathcal{W}(v) \subset \mathcal{W}$ and $\mathcal{V}(w) \subset \mathcal{V}$. We call the sets $\mathcal{V}(w)$ attribute cliques (each such set induces a clique in $\mathcal{G}$). For $v \in \mathcal{V}$ define

$$\mathcal{W}'(v) = \{w \in \mathcal{W}(v) : |\mathcal{V}(w)| \geq 2\}.$$

Furthermore, define the vertex sets SMALL $= \{v : |\mathcal{W}'(v)| \leq 0.1 \ln n\}$ and LARGE $= \mathcal{V} \setminus$ SMALL. Vertices belonging to SMALL (LARGE) we call small (large). Large vertices are those that have neighbours in many attribute cliques. While small vertices' neighbours are included in few attribute cliques (possibly even in only one).

Let

$$d_0 = mp(1 - (1 - p)^{n-1}), \qquad d_1 = nmp^2, \qquad k_0 = \max\{2, np\} \frac{\ln n}{\ln \ln n}.$$

Observe that $d_0 = \mathbb{E}|\mathcal{W}'(v)|$, d_1 is approximate expected degree of $\mathcal{G}(n, m, p)$. Note that (1) and $np = \Theta(1)$ imply

$$d_0 \asymp d_1 \asymp \ln n. \tag{3}$$

Let $w \in \mathcal{W}$, $\mathcal{S} \subset \mathcal{V}$, $|\mathcal{S}| = s$, and $\mathcal{V}_{\mathcal{S}}(w) = \mathcal{V}(w) \cap \mathcal{S}$. Than the number of edges included in both the attribute clique $\mathcal{V}(w)$ and $\mathcal{S}$ is $\binom{|\mathcal{V}_{\mathcal{S}}(w)|}{2}$ where $|\mathcal{V}_{\mathcal{S}}(w)|$ is binomially distributed $\mathrm{Bin}(s, p)$. Similarly the number of edges between $\mathcal{S}$ and $\bar{\mathcal{S}}$ provided by the attribute clique $\mathcal{V}(w)$ is $|\mathcal{V}_{\mathcal{S}}(w)| \cdot |\mathcal{V}_{\bar{\mathcal{S}}}(w)|$, where $|\mathcal{V}_{\mathcal{S}}(w)|$ and $|\mathcal{V}_{\bar{\mathcal{S}}}(w)|$ are, resp., binomially distributed $\mathrm{Bin}(s, p)$ and $\mathrm{Bin}(n - s, p)$. We might expect that not many edges are included in multiple attribute cliques thus it is reasonable to approximate $2e(\mathcal{S}, \mathcal{S})$ and $e(\mathcal{S}, \bar{\mathcal{S}})$ by the sums

$$2 \sum_{w \in \mathcal{W}} \binom{|\mathcal{V}_\mathcal{S}(w)|}{2} \quad \text{and} \quad \sum_{w \in \mathcal{W}} |\mathcal{V}_\mathcal{S}(w)| \cdot |\mathcal{V}_{\bar{\mathcal{S}}}(w)|, \quad \text{respectively.} \tag{4}$$

In the proofs we will need tight bounds for the above sums with exponential probabilities. At some point we will concentrate on those attributes for which $2 \leq |\mathcal{V}_\mathcal{S}(w)| \leq \ln \ln n$ (for the first sum) or $1 \leq |\mathcal{V}_\mathcal{S}(w)| \leq \ln \ln n$ and $1 \leq |\mathcal{V}_{\bar{\mathcal{S}}}(w)| \leq \ln \ln n$ (for the second sum). Therefore we introduce some additional notation.

Let $\eta_s \sim Bin(s, p)$ be a binomial random variable. For $1 \leq s \leq n - 1$, let

$$p_1 = p_1(s) = \Pr\{2 \leq \eta_s \leq \ln \ln n\},$$
$$p_2 = p_2(s) = \Pr\{1 \leq \eta_s \leq \ln \ln n\} \Pr\{1 \leq \eta_{n-s} \leq \ln \ln n\},$$
$$f_1 = f_1(s) = p_1(s)/(sp)^2, \quad f_2 = f_2(s) = p_2(s)/(s(n-s)p^2),$$
$$m_1 = m_1(s) = 0.5 f_1 s d_1, \quad m_2 = m_2(s) = f_2 s(n-s) m p^2.$$

Let $M_i = M_i(s) \sim Bin(m, p_i(s))$, $i = 1, 2$ be binomial random variables. Note that, given $\mathcal{S} \subset \mathcal{V}$ with $|\mathcal{S}| = s$, M_1 is the number of attributes $w \in \mathcal{W}$ with $2 \leq |\mathcal{V}_\mathcal{S}(w)| \leq \ln \ln n$ and M_2 counts those attributes for which both $\mathcal{V}_\mathcal{S}(w)$ and $\mathcal{V}_{\bar{\mathcal{S}}}(w)$ have sizes between 1 and $\ln \ln n$. Let $\{X_{i,j}, j \geq 1\}$, $i = 0, 1, 2$ be sequences of independent random variables with the distributions

$$\Pr\{X_{0,j} = k\} = \Pr\{\eta_s = k | 2 \leq \eta_s \leq \ln \ln n\}, \quad k = 2, 3, \ldots,$$
$$\Pr\{X_{1,j} = k\} = \Pr\{\eta_s = k | 1 \leq \eta_s \leq \ln \ln n\}, \quad k = 1, 2, \ldots,$$
$$\Pr\{X_{2,j} = k\} = \Pr\{\eta_{n-s} = k | 1 \leq \eta_{n-s} \leq \ln \ln n\}, \quad k = 1, 2, \ldots.$$

We will concentrate on analysing random variables

$$Y_1 = Y_1(s) = \sum_{1 \leq j \leq 2m_1} X_{0,j}(X_{0,j} - 1),$$
$$Y_2 = Y_2(s) = \sum_{1 \leq j \leq \lceil m_2/2 \rceil} X_{1,j} X_{2,j}.$$

We will show that Y_1 and Y_2 have approximately the same value as the random variables from (4) i.e. they are a good approximation of $2e(\mathcal{S}, \mathcal{S})$ and $e(\mathcal{S}, \bar{\mathcal{S}})$.

3   Proof of Theorem 1

Before we proceed with the proof of the main theorem we show two auxiliary lemmas.

Lemma 1. *Let $n, m \to +\infty$ and $p = p(n) \to 0$. Under conditions of Theorem 1 the following events hold with high probability*

(i) any two small vertices are at distance at least 3;
(ii) $\max_{w \in \mathcal{W}} |\mathcal{V}(w)| \le k_0$;
(iii) there is no vertex pair $\{u, v\} \subset \mathcal{V}$ belonging to three distinct attribute cliques, the number of pairs $\{u, v\}$ that belong to two attribute cliques is less than $\ln^3 n$.

Proof. (i) and (ii) follow from property P7 and Fact 10 of [2]. Let us show (iii). Let N_2 (N_3) be the number of pairs $\{u, v\} \in \mathcal{V}$ belonging to at least two (three) attribute cliques. We have for $i = 1, 2$

$$\mathbb{E}N_{1+i} \le \binom{n}{2}\binom{m}{i+1}p^{2+2i} \le (np)^2(mp)^2(mp^2)^{i-1} = \Theta(1)\ln^2 n\big(n^{-1}\ln n\big)^{i-1}.$$

By Markov's inequality, $\Pr\{N_3 \ge 1\} = O(n^{-1}\ln^3 n)$ and $\Pr\{N_2 > \ln^3 n\} = O(\ln^{-1} n)$. This finishes the proof of (iii).

Lemma 2. *Assume that conditions of Theorem 1 are satisfied. Then there exist constants $A_1, A_2, A_3, A_4 > 0$ such that uniformly in $s \in [n/d_1^3, n/2]$ and $n \ge A_4$, $m \ge A_4$ we have*

$$A_1 < f_i(s) < A_2, \quad i = 1, 2, \tag{5}$$
$$\Pr\{M_1(s) \ge 2m_1(s)\} \le \exp\left(-A_3\, s\ln n\right), \tag{6}$$
$$\Pr\{M_2(s) \le 0.5m_2(s)\} \le \exp\left(-A_3\, s\ln n\right), \tag{7}$$
$$\Pr\{Y_1(s) \ge 2sd_1\} \le \exp\left(-A_3 s(\ln n)/(\ln\ln n)^4\right), \tag{8}$$
$$\Pr\{Y_2(s) \le s(n - s)mp^2/4\} \le \exp\left(-A_3 s(\ln n)/(\ln\ln n)^4\right). \tag{9}$$

Proof. We prove (5) for $i = 1$. The proof for $i = 2$ is much the same.

Let $\hat{p}_1 = \Pr\{2 \le \Lambda \le \ln\ln n\}$, where $\Lambda \sim \mathcal{P}(sp)$ is a Poisson random variable with $\mathbb{E}\Lambda = sp$. The Poisson approximation error bound (Le Cam's lemma [19]) implies $|p_1 - \hat{p}_1| \le 2sp^2$. Hence it suffices to show (5) for $\hat{f}_1 = \hat{p}_1/(sp)^2$. We write

$$\hat{p}_1 = \hat{p}_{1.1} - \hat{p}_{1.2}, \quad \hat{p}_{1.1} = 1 - \Pr\{\Lambda \le 1\}, \quad \hat{p}_{1.2} = \Pr\{\Lambda > \ln\ln n\}$$

and easily verify (5) for $\hat{f}_{1.1} := \hat{p}_{1.1}/(sp)^2 = (1 - e^{-sp} - spe^{-sp})/(sp)^2$. Finally, we upper bound the Poisson tail probability (Proposition 1 of [7])

$$\hat{p}_{1.2} \le \left(1 - \left(\frac{sp}{\lceil\ln\ln n\rceil}\right)\right)^{-1} e^{-sp}\frac{(sp)^{\lceil\ln\ln n\rceil}}{\lceil\ln\ln n\rceil!} \le 2\frac{(sp)^{\lceil\ln\ln n\rceil}}{\lceil\ln\ln n\rceil!}$$

and show that $\hat{p}_{1.2}/(sp)^2 \le 2(c_2/2)^{\lceil\ln\ln n\rceil - 2}/(\lceil\ln\ln n\rceil!) \le \ln^{-1} n$. This finishes the proof of (5).

Proof of (6), (7). We have

$$\mathbb{E}M_1 = mp_1 \le 0.5mf_1snp^2 = 0.5f_1d_1s = m_1,$$
$$\mathbb{E}M_2 = mp_2 = m_2.$$

Chernoff's bounds (see Theorem 2.1 in [10]) imply

$$\Pr\{M_1 \ge 2m_1\} \le \Pr\{M_1 \ge mp_1 + m_1\}$$
$$\le \exp\left(-\frac{m_1^2}{2(mp_1 + m_1/3)}\right) \le \exp\left(-(3/8)m_1\right),$$
$$\Pr\{M_2 \le 0.5m_2\} \le \exp\left(-m_2/8\right).$$

Now (6), (7) follow from (3), (5).

Proof of (8), (9). We have

$$\mathbb{E}Y_1 = 2m_1\mathbb{E}(X_{0,1}(X_{0,1} - 1)) \le 2m_1p_2^{-1}\mathbb{E}\eta_s(\eta_s - 1) = 2m_1p_2^{-1}s(s-1)p^2 \le sd_1,$$
$$\mathbb{E}Y_2 \ge \frac{m_2}{2}(\mathbb{E}X_{1,1})(\mathbb{E}X_{2,1}) \ge \frac{m_2s(n-s)p^2}{3p_2} = \frac{1}{3}s(n-s)mp^2.$$

Here we used $(\mathbb{E}X_{1,1})(\mathbb{E}X_{2,1}) = p_2^{-1}(\mathbb{E}\eta_s\mathbb{I}_{\{\eta_s \le \ln\ln n\}})(\mathbb{E}\eta_{n-s}\mathbb{I}_{\{\eta_{n-s} \le \ln\ln n\}})$ and the fact that $\mathbb{E}\eta_i\mathbb{I}_{\{\eta_i \le \ln\ln n\}} = (1 + o(1))\mathbb{E}\eta_i$ uniformly in $n/d_1^3 \le i \le n$ as $n \to +\infty$. Next we apply Hoeffding's inequality [9] to the sums Y_1, Y_2 with summands bounded by $(\ln\ln n)^2$. We have

$$\Pr\{Y_1 \ge 2sd_1\} \le \Pr\{Y_1 \ge \mathbb{E}Y_1 + sd_1\} \le \exp\left(-\frac{s^2d_1^2}{m_1(\ln\ln n)^4}\right),$$
$$\Pr\left\{Y_2 < \frac{1}{4}s(n-s)mp^2\right\} \le \exp\left(-2\frac{(s(n-s)mp^2/12)^2}{\lceil m_2/2\rceil(\ln\ln n)^4}\right).$$

Now (8), (9) follow from (3), (5). This finishes the proof of Lemma 2.

Now we proceed with the proof of Theorem 1.

Let $\varphi(\mathcal{S}) = \frac{e(\mathcal{S},\bar{\mathcal{S}})}{2e(\mathcal{S},\mathcal{S})+e(\mathcal{S},\bar{\mathcal{S}})}$. Then $\Phi(\mathcal{G}) = \min_{\mathcal{S}\subset\mathcal{V},|\mathcal{S}|\le n/2}\varphi(\mathcal{S})$. Note that graph $\mathcal{G}(n,m,p)$ is connected with high probability, [15,17]. Hence with high probability $\varphi(\{v\}) = 1, \forall v \in \mathcal{V}$. We show below that for some $\{\varepsilon_n\}$ such that $\varepsilon_n \to 0$ as $n \to \infty$,

$$|\varphi(\mathcal{S}) - 1| \le \varepsilon_n \quad \forall\, \mathcal{S} \subset \mathcal{V}, \quad 2 \le |\mathcal{S}| \le n/d_1^3 \tag{10}$$

with high probability. Furthermore, for some $\{\varepsilon_n'\}$ such that $\varepsilon_n' \to 0$ as $n \to \infty$, inequalities

$$e(\mathcal{S},\bar{\mathcal{S}}) \ge (1 + \varepsilon_n')|\mathcal{S}|d_1/8, \quad e(\mathcal{S},\mathcal{S}) \le (1 + \varepsilon_n')|\mathcal{S}|d_1 \tag{11}$$

hold uniformly in $\mathcal{S} \subset \mathcal{V}$, $n/d_1^3 \le |\mathcal{S}| \le n/2$, with high probability. From (10) and (11) we derive (2). It remains to prove (10) and (11).

Proof of (10). Given $\mathcal{S} \subseteq \mathcal{V}$ let

$$W^k(\mathcal{S}) = |\{w \in \mathcal{W} : |\mathcal{V}_\mathcal{S}(w)| \geq k\}|, \qquad \mathcal{V}_\mathcal{S}(w) = \mathcal{V}(w) \cap \mathcal{S}. \qquad (12)$$

We claim that the event

$$W^k(\mathcal{S}) \leq 4k^{-1}|\mathcal{S}|, \text{ for every } 2 \leq k \leq nd_1^{-3} \text{ and } k \leq |\mathcal{S}| \leq nd_1^{-3} \qquad (13)$$

has probability $1 - o(1)$. Indeed, by the union bound, the probability of the complement event is at most

$$\sum_{k=2}^{n/d_1^3} \sum_{s=k}^{n/d_1^3} \binom{n}{s} \binom{m}{\frac{4s}{k}} \left(\binom{s}{k} p^k \right)^{4s/k}$$

$$\leq \sum_{k=2}^{n/d_1^3} \sum_{s=k}^{n/d_1^3} \left(\frac{en}{s} \cdot \frac{e^{4/k} m^{4/k} k^{4/k}}{4^{4/k} s^{4/k}} \cdot \frac{e^4 s^4 p^4}{k^4} \right)^s$$

$$= \sum_{k=2}^{n/d_1^3} \sum_{s=k}^{n/d_1^3} \left(\frac{e^{5+(4/k)}}{k^{4-(4/k)}} \cdot \frac{s}{n} \cdot \left(\frac{s}{m} \right)^{2--4/k} (nmp^2)^2 \right)^s$$

$$\leq \sum_{k=2}^{n/d_1^3} \left(\frac{A_5}{d_1} \right)^k = o(1).$$

Here A_5 is a constant independent of k and n. In the second inequality we used $\binom{t}{u} \leq (et/u)^u$. In the last inequality we used $s/n \leq d_1^{-3}$ and $s/m \leq 1$.

Now we show that for some sequence $\{\varepsilon'_n\}$, such that $\varepsilon'_n \to 0$ as $n \to \infty$, we have with high probability

$$e(\mathcal{S}, \mathcal{S}) \leq \varepsilon'_n d_1 |\mathcal{S}| \quad \forall \, \mathcal{S} \subset \mathcal{V}, \quad 2 \leq |\mathcal{S}| \leq d_1^{-3} n. \qquad (14)$$

Given $0 < \varepsilon < 1$ let

$$I = ((\ln(\ln k_0 - \ln 2) - \ln \ln \varepsilon^{-1})/\ln 2) - 1.$$

Note that $\varepsilon^{2^{I+1}} k_0 = 2$ and $I = O(\ln \ln \ln n)$. Given realised instance of $\mathcal{G}(n, m, p)$ satisfying Lemma 1 (ii) and (13) we have for any $\mathcal{S} \subseteq \mathcal{V}$ with $|\mathcal{S}| \leq n/d_1^3$

$$2e(\mathcal{S}, \mathcal{S}) \leq \sum_{w \in \mathcal{W} : |\mathcal{V}_\mathcal{S}(w)| \geq 2} \binom{|\mathcal{V}_\mathcal{S}(w)|}{2}$$

$$\leq \sum_{i=0}^{I} \sum_{w \in \mathcal{W} : \varepsilon^{2^{i+1}} k_0 \leq |\mathcal{V}_\mathcal{S}(w)| \leq \varepsilon^{2^i} k_0} \left(\varepsilon^{2^i} k_0 \right)^2 + \sum_{w \in \mathcal{W} : \varepsilon k_0 \leq |\mathcal{V}_\mathcal{S}(w)| \leq k_0} k_0^2$$

$$\leq \sum_{i=0}^{I} W^{\varepsilon^{2^{i+1}} k_0}(\mathcal{S}) \varepsilon^{2^{i+1}} k_0^2 + W^{\varepsilon k_0}(\mathcal{S}) k_0^2$$

$$\leq \sum_{i=0}^{I} \frac{4|\mathcal{S}|}{\varepsilon^{2^{i+1}} k_0} \varepsilon^{2^{i+1}} k_0^2 + \frac{4|\mathcal{S}|}{\varepsilon k_0} k_0^2$$

$$\leq 4|\mathcal{S}| k_0 (I + 1 + \varepsilon^{-1}).$$

The last inequality combined with (3) shows (14).

Next we lowerbound $e(\mathcal{S}, \bar{\mathcal{S}})$. For $\mathcal{S} \subset \text{LARGE}$ let $\mathcal{W}_{\mathcal{S},\bar{\mathcal{S}}}$ be the set of triples (u, v, w) where $u \in \mathcal{S}$, $v \in \bar{\mathcal{S}}$, $w \in \mathcal{W}$ and $u, v \in \mathcal{V}(w)$. Clearly, $|\mathcal{W}_{\mathcal{S},\bar{\mathcal{S}}}| \geq e(\mathcal{S}, \bar{\mathcal{S}})$. Observe that Lemma 1 (iii) implies $2e(\mathcal{S}, \bar{\mathcal{S}}) \geq |\mathcal{W}_{\mathcal{S},\bar{\mathcal{S}}}|$ as each edge may belong to at most two distinct attribute cliques. Furthermore, we have

$$|\mathcal{W}_{\mathcal{S},\bar{\mathcal{S}}}| = \sum_{u \in \mathcal{S}} \sum_{w \in \mathcal{W}'(u)} |\mathcal{V}_{\bar{\mathcal{S}}}(w)| \geq \sum_{u \in \mathcal{S}} \sum_{w \in \mathcal{W}'(u)} \mathbb{I}_{\{\mathcal{V}_{\bar{\mathcal{S}}}(w) \neq \emptyset\}}$$

$$\geq \sum_{u \in \mathcal{S}} |\mathcal{W}'(u)| - \sum_{w \in \mathcal{W}: \mathcal{V}(w) \subset \mathcal{S}, |\mathcal{V}(w)| \geq 2} |\mathcal{V}(w)|. \tag{15}$$

By Lemma 1 (ii) and (13), the quantity (15) is at least

$$|\mathcal{S}|0.1 \ln n - W^2(\mathcal{S})k_0 \geq |\mathcal{S}|0.1 \ln n - 2|\mathcal{S}|k_0 \geq 0.09|\mathcal{S}| \ln n.$$

Invoking (3) we obtain for some constant A_6 that with high probability

$$e(\mathcal{S}, \bar{\mathcal{S}}) \geq |\mathcal{W}_{\mathcal{S},\bar{\mathcal{S}}}|/2 \geq 0.04|\mathcal{S}| \ln n \geq A_6|\mathcal{S}|d_1, \quad \forall_{\mathcal{S} \subset \text{LARGE}}. \tag{16}$$

Now we show (10). Given $\mathcal{S} \subseteq \mathcal{V}$ with $|\mathcal{S}| \leq n/d_1^3$, let $\mathcal{S}_1 = \mathcal{S} \cap \text{SMALL}$ and $\mathcal{S}_2 = \mathcal{S} \cap \text{LARGE}$. Lemma 1 (i) implies $e(\mathcal{S}_1, \mathcal{S}_1) = 0$ and $e(\mathcal{S}_1, \mathcal{S}_2) \leq |\mathcal{S}_2|$. For $\mathcal{S}_2 = \emptyset$ we obtain $\varphi(\mathcal{S}) = \varphi(\mathcal{S}_1)$. But $\varphi(\mathcal{S}_1) = 1$ with high probability because $\mathcal{G}(n, m, p)$ is connected with high probability. For $\mathcal{S}_2 \neq \emptyset$ we have

$$e(\mathcal{S}, \mathcal{S}) = e(\mathcal{S}_1, \mathcal{S}_1) + e(\mathcal{S}_2, \mathcal{S}_2) + e(\mathcal{S}_1, \mathcal{S}_2) = e(\mathcal{S}_2, \mathcal{S}_2) + e(\mathcal{S}_1, \mathcal{S}_2),$$

$$e(\mathcal{S}, \bar{\mathcal{S}}) = e(\mathcal{S}_1, \bar{\mathcal{S}}) + e(\mathcal{S}_2, \bar{\mathcal{S}}) \geq e(\mathcal{S}_2, \bar{\mathcal{S}}) = e(\mathcal{S}_2, \bar{\mathcal{S}}_2) - e(\mathcal{S}_1, \mathcal{S}_2).$$

Furthermore,

$$e(\mathcal{S}, \mathcal{S}) \leq e(\mathcal{S}_2, \mathcal{S}_2) + |\mathcal{S}_2|,$$

$$2e(\mathcal{S}, \mathcal{S}) + e(\mathcal{S}, \bar{\mathcal{S}}) \geq 2e(\mathcal{S}_2, \mathcal{S}_2) + e(\mathcal{S}_2, \bar{\mathcal{S}}_2).$$

The latter inequalities imply

$$0 \leq 1 - \varphi(\mathcal{S}) = \frac{2e(\mathcal{S}, \mathcal{S})}{2e(\mathcal{S}, \mathcal{S}) + e(\mathcal{S}, \bar{\mathcal{S}})} \leq \frac{2 + 2|\mathcal{S}_2|^{-1}e(\mathcal{S}_2, \mathcal{S}_2)}{2|\mathcal{S}_2|^{-1}e(\mathcal{S}_2, \mathcal{S}_2) + |\mathcal{S}_2|^{-1}e(\mathcal{S}_2, \bar{\mathcal{S}}_2)}.$$

By (14), (16), the right side is $o(1)$ uniformly in $\mathcal{S} \subset \mathcal{V}$, $|\mathcal{S}| \leq n/d_1^3$ with high probability. This finishes the proof of (10).

Proof of (11). Let us prove the second part of (11). Let $\mathcal{W}_1 = \{w : |\mathcal{V}(w)| \geq \ln \ln n\}$ and $\mathcal{W}_2 = \mathcal{W} \setminus \mathcal{W}_1$. For any $\mathcal{S} \subset \mathcal{V}$ we have

$$2e(\mathcal{S}, \mathcal{S}) \leq \sum_{w \in \mathcal{W}} |\mathcal{V}_{\mathcal{S}}(w)| \cdot \big(|\mathcal{V}_{\mathcal{S}}(w)| - 1\big) = \tilde{e}_1(\mathcal{S}, \mathcal{S}) + \tilde{e}_2(\mathcal{S}, \mathcal{S}). \tag{17}$$

Here $\tilde{e}_1(\mathcal{S}, \mathcal{S})$ and $\tilde{e}_2(\mathcal{S}, \mathcal{S})$ stand for the sums over $w \in \mathcal{W}_1$ and $w \in \mathcal{W}_2$ respectively. Note that

$$\tilde{e}_1(\mathcal{S}, \mathcal{S}) \leq \tilde{e}_1(\mathcal{V}, \mathcal{V}) = \sum_{w \in \mathcal{W}_1} |\mathcal{V}(w)| \cdot \big(|\mathcal{V}(w)| - 1\big). \tag{18}$$

Furthermore,

$$\mathbb{E}\left(\sum_{w \in \mathcal{W}_1} |\mathcal{V}(w)|(|\mathcal{V}(w)| - 1)\right) = m \sum_{t \geq \ln\ln n} \binom{n}{t} p^t (1-p)^{n-t} t(t-1)$$

$$\leq m \sum_{t \geq \ln\ln n} \left(\frac{enp}{t}\right)^t t^2$$

$$\leq e^2 n^2 m p^2 \sum_{t \geq \ln\ln n} \left(\frac{enp}{\ln\ln n}\right)^{t-2}$$

$$\leq e^2 n^2 m p^2 (1 + o(1)) \left(\frac{enp}{\ln\ln n}\right)^{\ln\ln n - 2}$$

$$\leq n^2 m p^2 \ln^{-9} n.$$

In the last step we used $np = \Theta(1)$ and assumed that n is large. Hence we get $\mathbb{E}\tilde{e}_1(\mathcal{V}, \mathcal{V}) \leq nd_1 \ln^{-9} n$. From (18) we obtain, by Markov's inequality and (3),

$$\Pr\left\{\max_{\mathcal{S}, |\mathcal{S}| \geq n/d_1^3} \frac{\tilde{e}_1(\mathcal{S}, \mathcal{S})}{|\mathcal{S}|} \geq d_1^{-1}\right\} \leq \Pr\left\{\tilde{e}_1(\mathcal{V}, \mathcal{V}) \geq n/d_1^4\right\} \leq d_1^5 \ln^{-9} n = o(1). \tag{19}$$

Next we show that the event

$$\tilde{e}_2(\mathcal{S}, \mathcal{S}) \leq 2|\mathcal{S}|d_1 \quad \forall \, \mathcal{S} \subset \mathcal{V}, \quad n/d_1^3 \leq |\mathcal{S}| \leq n/2 \tag{20}$$

holds with high probability. Given $\mathcal{S}$, let $M_{\mathcal{S}}$ be the number of attributes $w \in \mathcal{W}$ such that $2 \leq |\mathcal{V}_{\mathcal{S}}(w)| \leq \ln\ln n$. Let $\sum_{\mathcal{S}}$ stand for the sum over $\mathcal{S} \subset \mathcal{V}$ with $n/d_1^3 \leq |\mathcal{S}| \leq n/2$. By the union bound, the probability of the event complement to (20) is at most

$$\sum_{\mathcal{S}} \Pr\left\{\tilde{e}_2(\mathcal{S}, \mathcal{S}) > 2|\mathcal{S}|d_1\right\} \tag{21}$$

$$\leq \sum_{\mathcal{S}} \Pr\left\{\tilde{e}_2(\mathcal{S}, \mathcal{S}) > 2|\mathcal{S}|d_1 \,\Big|\, M_{\mathcal{S}} \leq 2m_1(|\mathcal{S}|)\right\} + \sum_{\mathcal{S}} \Pr\left\{M_{\mathcal{S}} \geq 2m_1(|\mathcal{S}|)\right\}$$

$$\leq \sum_{s=n/d_1^3}^{n/2} \binom{n}{s} \Pr\left\{Y_1(s) \geq 2sd_1\right\} + \sum_{s=n/d_1^3}^{n/2} \binom{n}{s} \Pr\left\{M_1(s) \geq 2m_1(s)\right\}$$

$$\leq \sum_{s=n/d_1^3}^{n/2} \exp\left(s \ln\frac{n}{s} + s - A_3 \frac{s \ln n}{(\ln\ln n)^4}\right)$$

$$+ \sum_{s=n/d_1^3}^{n/2} \exp\left(s \ln\frac{n}{s} + s - A_3 s \ln n\right)$$

$$= o(1).$$

Here we used (6), (8), the inequality $\ln\binom{n}{s} \leq s + s\ln(n/s)$ and (3). The second part of (11) follows from (17), (19), (20).

Let us prove the first part of (11). Let

$$T(\mathcal{S}) = \sum_{w \in \mathcal{W}} |\mathcal{V}_{\mathcal{S}}(w)||\mathcal{V}_{\bar{\mathcal{S}}}(w)| \mathbb{I}_{\{1 \leq |\mathcal{V}_{\mathcal{S}}(w)| \leq \ln \ln n\}} \mathbb{I}_{\{1 \leq |\mathcal{V}_{\bar{\mathcal{S}}}(w)| \leq \ln \ln n\}}.$$

Lemma 1 (iii) implies that with high probability $e(\mathcal{S}, \bar{\mathcal{S}}) \geq T(\mathcal{S}) - \ln^3 n$. To prove (11) we show that the event

$$T(\mathcal{S}) \geq |\mathcal{S}|(n - |\mathcal{S}|)mp^2/4 \quad \forall \mathcal{S} \subset \mathcal{V}, \quad n/d_1^3 \leq |\mathcal{S}| \leq n/2 \tag{22}$$

holds with high probability. Given $\mathcal{S} \subset \mathcal{V}$, let $\tilde{M}_{\mathcal{S}}$ be the number of attributes $w \in \mathcal{W}$ such that $1 \leq |\mathcal{V}_{\mathcal{S}}(w)| \leq \ln \ln n$ and $1 \leq |\mathcal{V}_{\bar{\mathcal{S}}}(w)| \leq \ln \ln n$. Let $\sum_{\mathcal{S}}$ stand for the sum over $\mathcal{S} \subset \mathcal{V}$, $n/d_1^3 \leq |\mathcal{S}| \leq n/2$. By the union bound, the probability of the event complement to (22) is at most

$$\sum_{\mathcal{S}} \Pr\left\{T(\mathcal{S}) \leq |\mathcal{S}|(n - |\mathcal{S}|)mp^2/4\right\}$$

$$\leq \sum_{\mathcal{S}} \Pr\left\{T(\mathcal{S}) \leq |\mathcal{S}|(n - |\mathcal{S}|)mp^2/4 \,\middle|\, \tilde{M}_{\mathcal{S}} \geq 0.5m_2(|\mathcal{S}|)\right\}$$

$$+ \sum_{\mathcal{S}} \Pr\left\{\tilde{M}_{\mathcal{S}} \leq 0.5m_2(|\mathcal{S}|)\right\}$$

$$\leq \sum_{s=n/d_1^3}^{n/2} \binom{n}{s} \Pr\left\{Y_2(s) \leq s(n - s)mp^2/4\right\}$$

$$+ \sum_{s=n/d_1^3}^{n/2} \binom{n}{s} \Pr\left\{M_2(s) \leq 0.5m_2(s)\right\}$$

$$\leq \sum_{s=n/d_1^3}^{n/2} \exp\left(s \ln \frac{n}{s} + s - A_3 s \frac{\ln n}{(\ln \ln n)^4}\right)$$

$$+ \sum_{s=n/d_1^3}^{n/2} \exp\left(s \ln \frac{n}{s} + s - A_3 s \ln n\right)$$

$$= o(1).$$

This completes the proof of Theorem 1.

References

1. Bloznelis, M., Godehardt, E., Jaworski, J., Kurauskas, V., Rybarczyk, K.: Recent progress in complex network analysis: properties of random intersection graphs. In: Lausen, B., Krolak-Schwerdt, S., Böhmer, M. (eds.) Data Science, Learning by Latent Structures, and Knowledge Discovery. SCDAKO, pp. 79–88. Springer, Heidelberg (2015). https://doi.org/10.1007/978-3-662-44983-7_7

2. Bloznelis, M., Jaworski, J., Rybarczyk, K.: The cover time of a sparse random intersection graph. arXiv:1910.09639 (2019)

3. Britton, T., Deijfen, M., Lagerås, A.N., Lindholm, M.: Epidemics on random graphs with tunable clustering. J. Appl. Probab. **45**(3), 743–756 (2008)

4. Deijfen, M., Kets, W.: Random intersection graphs with tunable degree distribution and clustering. Probab. Engrg. Inform. Sci. **23**, 661–674 (2009)

5. Eschenauer, L, Gligor, V.D.: A key-management scheme for distributed sensor networks. In: Proceedings of the 9th ACM Conference on Computer and Communications Security, pp. 41–47 (2002)

6. Frieze, A., Karonski, M.: Introduction to Random Graphs. Cambridge University Press, Cambridge (2016)

7. Glynn, P.W.: Upper bounds for Poisson tail probabilities. Oper. Res. Lett. **6**, 9–14 (1987)

8. Godehardt, E., Jaworski, J.: Two models of random intersection graphs for classification. In: Schwaiger, M., Opitz, O. (eds.) Exploratory Data Analysis in Empirical Research. Studies in Classification, Data Analysis, and Knowledge Organization, pp. 67–81. Springer, Heidelberg (2003). https://doi.org/10.1007/978-3-642-55721-7_8

9. Hoeffding, W.: Probability inequalities for sums of bounded random variables. J. Amer. Stat. Assoc. **58**, 13–30 (1963)

10. Janson, S., Łuczak, T., Ruciński, A.: Random Graphs. Wiley, New York (2001)

11. Karoński, M., Scheinerman, E.R., Singer-Cohen, K.B.: On random intersection graphs: the subgraph problem. Comb. Probab. Comput. **8**, 131–159 (1999)

12. Newman, M.E.J., Strogatz, S.H., Watts, D.J.: Random graphs with arbitrary degree distributions and their applications. Phys. Rev. E **64**, 026118 (2002)

13. Nikoletseas, S.E., Raptopoulos, C., Spirakis, P.G.: Expander properties and the cover time of random intersection graphs. In: Kučera, L., Kučera, A. (eds.) MFCS 2007. LNCS, vol. 4708, pp. 44–55. Springer, Heidelberg (2007). https://doi.org/10.1007/978-3-540-74456-6_6

14. Nikoletseas, S., Raptopoulos, C., Spirakis, P.G.: The second eigenvalue of random walks on symmetric random intersection graphs. In: Bozapalidis, S., Rahonis, G. (eds.) CAI 2007. LNCS, vol. 4728, pp. 236–246. Springer, Heidelberg (2007). https://doi.org/10.1007/978-3-540-75414-5_15

15. Rybarczyk, K.: The coupling method for inhomogeneous random intersection graphs. Electron. J. Combin. **24**(2) (2017). Paper 2.10

16. Sinclair, A., Jerrum, M.: Approximate counting, uniform generation and rapidly mixing Markov chains. Inf. Comput. **82**, 93–133 (1989)

17. Singer-Cohen, K.B.: Random Intersection Graphs. Ph.D. thesis, The Johns Hopkins University, Department of Mathematical Sciences (1995)

18. Spirakis, P.G., Nikoletseas, S., Raptopoulos, C.: A guided tour in random intersection graphs. In: Fomin, F.V., Freivalds, R., Kwiatkowska, M., Peleg, D. (eds.) ICALP 2013. LNCS, vol. 7966, pp. 29–35. Springer, Heidelberg (2013). https://doi.org/10.1007/978-3-642-39212-2_5

19. Steele, J.M.: Le Cam's inequality and Poisson approximations. Am. Math. Monthly **101**, 48–54 (1994)

Iterated Global Models
for Complex Networks

Anthony Bonato$^{(\boxtimes)}$ and Erin Meger

Ryerson University, Toronto, ON, Canada
{abonato,erin.k.meger}@ryerson.ca

Abstract. We introduce the Iterated Global model as a deterministic graph process that simulates several properties of complex networks. In this model, for every set S of nodes of a prescribed cardinality, we add a new node that is adjacent to every node in S. We focus on the case where the size of S is approximately half the number of nodes at each time-step, and we refer to this as the half-model. The half-model provably generate graphs that densify over time, have bad spectral expansion, and low diameter. We derive the clique, chromatic, and domination numbers of graphs generated by the model.

Keywords: Network models · Social networks · Densification · Spectral Graph Theory

1 Introduction

Over the last two decades, research in modelling complex networks has become of great interest to mathematicians and theoretical computer scientists. Complex networks arise in technological, social, and biological contexts. The emergence of the study of complex networks such as the web graph and on-line social networks has focused attention on these large-scale graphs, and in the modeling and mining of their emergent properties; see [1,5,6] for more on these models.

Two deterministic models of complex networks of particular interest to the current study were introduced: the *Iterated Local Transitivity (ILT)* model and the *Iterated Local Anti-Transitivity (ILAT)* model [3,4]. Consider a social network where friendships have positive edge signs and adversarial relations have negative edge signs. A *triad* is a set of three nodes in a signed network. A triad is said to be *balanced* if the product of the edge signs is positive. Structural balance theory states that these networks seek to balance all triads [8]. The ILT and ILAT models were designed with balanced triads in mind. In the ILT model, nodes are *cloned*, where nodes are adjacent to all neighbors of their parent node. In the ILAT model, nodes are *anti-cloned*, where a new node is adjacent to all non-neighbors of it's parent node. The ILT and ILAT models simulates many properties of social networks. For example, as shown in [4], graphs generated

The first author acknowledges funding from an NSERC Discovery grant.

B. Kamiński et al. (Eds.): WAW 2020, LNCS 12091, pp. 135–144, 2020.
https://doi.org/10.1007/978-3-030-48478-1_10

by the model densify over time (see [10] for more on densification), and exhibit bad spectral expansion (see [9] for more on this topic in social networks). In addition, the ILT model generates graphs with the small-world property, which requires the graphs to have low diameter and dense neighbor sets. Both the ILT and ILAT models were unified in the recent context of Iterated Local Models in [2].

The ILT, ILAT, and ILM models focused on considering the local structure of the graph and generating a new model iteratively from this structure. We now define a model that is independent of the structure of the initial graph but retains the iterative character of the previously defined models. We introduce the *Iterated Global Models*, where a dominating node is added for each subset of nodes of a given cardinality.

Let $k \geq 1$ be an integer. The one parameter of the model is the initial, connected graph $G = G_0$. At each time-step $t \geq 0$, we create G_{t+1} from G_t in the following way: for each set of nodes of cardinality $\lfloor \frac{1}{k}|V(G_t)| \rfloor$, say S, add a new v_S that is adjacent to each node of S. We name this process the $\frac{1}{k}$-*model*. For ease of notation, we refer to newly added nodes in G_{t+1} as *clones*. Note that the clones form an independent set in G_{t+1}.

For the sake of clarity, we focus in this paper on the case $k = 2$, which we refer to as the *half-model*. In the half-model, each new node is adjacent to approximately half of the existing network. See Fig. 1 for an example.

While structural balance theory considers the importance of local ties, the half-model may be useful in analyzing complex networks where nodes interact via weaker, non-local ties. In social networks such as Twitter, Instagram, or Reddit, we may form a network of users where links are determined by likes, comments, or comments. For example, a user on Reddit may choose to comment on a fraction of the posts they read, which is reflective of the design of the half-model.

The paper is organized as follows. In Sect. 2, we prove that, as observed in complex networks, the half-model densifies over time and has bad spectral expansion. We also show that after five time-steps, graphs generated by the

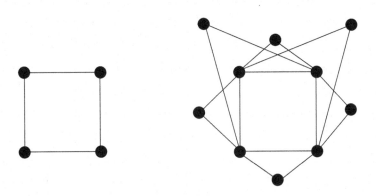

Fig. 1. One time-step of the half-model beginning with C_4.

model have diameter at most 3. The half-model is of graph theoretic interest in its own right, and in Sect. 3 we determine the clique, chromatic, and domination numbers of graphs generated by the model. We conclude with further directions to investigate for the half-model.

For a general reference on graph theory, the reader is directed to [13]. For background on social and complex networks, see [1,5,7]. Throughout the paper, we consider finite, undirected graphs.

2   Complex Network Properties of the Half-Model

Our first result establishes the order and size of graphs generated by the half-model. We first recall Stirling's approximation for the factorial given by

$$n! \sim \sqrt{2\pi n}\left(\frac{n}{e}\right)^n.$$

Stirling's approximation may be used to derive an expression for the central binomial coefficient given by

$$\binom{2n}{n} \sim \frac{2^{2n}}{\sqrt{\pi n}},$$

which may be derived directly and is part of folklore. Such an approximation will be useful in our analysis, and its usefulness has provided motivation for the study of the half-model as opposed to other values of k. For an exposition of the asymptotics of binomial coefficients, see the book [12].

The number of nodes of G_t is denoted by n_t, the number of edges is denoted by e_t.

Theorem 1. *The order and size of the graph G_t in the half-model are given by the following, respectively:*

$$n_t \sim \binom{n_{t-1}}{\lfloor \frac{n_{t-1}}{2} \rfloor} \qquad and \qquad e_t \sim \binom{n_{t-1}}{\lfloor \frac{n_{t-1}}{2} \rfloor} \cdot \left\lfloor \frac{n_{t-1}}{2} \right\rfloor.$$

Before we give the proof of Theorem 1, we simplify notation by defining the function

$$\alpha_t = \binom{n_t}{\lfloor \frac{n_t}{2} \rfloor}.$$

Proof. We begin with the order of G_t. By the definition of the model, at each time-step $t \geq 1$, we add one node for each set of size $\lfloor \frac{n_{t-1}}{2} \rfloor$. Hence, we derive the following sum given by

$$n_t = n_0 + \sum_{i=1}^{t} \alpha_{i-1}.$$

The term α_{t-1} will dominate the rest of the summation, which gives us the desired expression for the order of G_t.

Next, we determine the size of G_t. Each new node added is adjacent to a set of size $\lfloor \frac{n_{t-1}}{2} \rfloor$, and we add α_{t-1} nodes, so we obtain the following recursive formula for the number of edges at time-step t :

$$e_t = e_{t-1} + \left\lfloor \frac{n_{t-1}}{2} \right\rfloor \alpha_{t-1}.$$

We observe that the second term dominates the sum, and the result follows. □

We say that a network *densifies* if the limit of the ratio of edges to nodes is unbounded. Densification power laws in complex networks were first reported in [10]. From Theorem 1 we have the following result.

Corollary 1. *The half-model densifies with time.*

Proof. By Theorem 1, we have that

$$\frac{e_t}{n_t} \sim \frac{\alpha_{t-1} \cdot \left\lfloor \frac{n_{t-1}}{2} \right\rfloor}{\alpha_{t-1}} = \left\lfloor \frac{n_{t-1}}{2} \right\rfloor,$$

which tends to infinity with t. □

For a graph G and sets of nodes $X, Y \subseteq V(G)$, define $E(X, Y)$ to be the set of edges in G with one endpoint in X and the other in Y. For simplicity, we write $E(X) = E(X, X)$. Let A denote the adjacency matrix and D denote the diagonal degree matrix of a graph G. The *normalized Laplacian* of G is

$$\mathcal{L} = I - D^{-1/2} A D^{-1/2}.$$

Let $0 = \lambda_0 \le \lambda_1 \le \cdots \le \lambda_{n-1} \le 2$ denote the eigenvalues of $\mathcal{L}$. The *spectral gap* of the normalized Laplacian is defined as

$$\lambda = \max\{|\lambda_1 - 1|, |\lambda_{n-1} - 1|\}.$$

We will use the expander mixing lemma for the normalized Laplacian [6]. For sets of nodes X and Y, we use the notation $\mathrm{vol}(X) = \sum_{v \in X} \deg(v)$ for the volume of X, $\overline{X} = V \setminus X$ for the complement of X, and, $e(X, Y)$ for the number of edges with one end in each of X and Y. Note that $X \cap Y$ need not be empty, and in this case, the edges completely contained in $X \cap Y$ are counted twice. In particular, $e(X, X) = 2|E(X)|$.

Lemma 1 (Expander mixing lemma). *[6] If G is a graph with spectral gap λ, then, for all sets $X \subseteq V(G)$,*

$$\left| e(X, X) - \frac{(\mathrm{vol}(X))^2}{\mathrm{vol}(G)} \right| \le \lambda \frac{\mathrm{vol}(X) \mathrm{vol}(\overline{X})}{\mathrm{vol}(G)}.$$

A spectral gap bounded away from zero is an indication of bad expansion properties, which is characteristic for social networks, [9]. The next theorem represents a drastic departure from the good expansion found in binomial random graphs, where $\lambda = o(1)$ [6].

Theorem 2. *Graphs generated by the half-model satisfy $\lambda_t \sim 1$, where λ_t is the spectral gap of G_t.*

Proof. Let $X = V(G_t)\backslash V(G_{t-1})$ be the set of cloned nodes added to G_{t-1} to form G_t. Since X is an independent set, we note that $e(X, X) = 0$. We derive that

$$\mathrm{Vol}(G_t) = 2e_t \sim \alpha_{t-1} \cdot n_{t-1},$$

$$\mathrm{Vol}(X) = \alpha_{t-1} \cdot \left\lfloor \frac{n_{t-1}}{2} \right\rfloor,$$

$$\mathrm{Vol}(\overline{X}) \sim \alpha_{t-1} \cdot \left\lfloor \frac{n_{t-1}}{2} \right\rfloor.$$

Hence, by Lemma 1, we have that

$$
\begin{aligned}
\lambda_t &\geq \frac{(\mathrm{Vol}(X))^2}{\mathrm{Vol}(G_t)} \cdot \frac{\mathrm{Vol}(G_t)}{\mathrm{Vol}(X)\mathrm{Vol}(\overline{X})} \\
&= \frac{\mathrm{Vol}(X)}{\mathrm{Vol}(\overline{X})} \\
&\sim \frac{\alpha_{t-1} \cdot \left\lfloor \frac{n_{t-1}}{2} \right\rfloor}{\alpha_{t-1} \cdot \left\lfloor \frac{n_{t-1}}{2} \right\rfloor} \\
&= 1,
\end{aligned}
$$

and the result follows.   □

We observe that the half-model has a small (in fact, constant) diameter as required for the small-world property. We first prove some results about the connectivity for graphs generated by this model.

Lemma 2. *For all $t \geq 0$, if G_t is connected and $n_t \geq 2$, then G_{t+1} is connected.*

Proof. If v is a clone in G_{t+1}, then since $n_t \geq 2$, we have that v is adjacent to at least one node u in $V(G_t)\backslash V(G_{t+1})$. Since G_t is connected by hypothesis, there exists a path from u to any other node of G_t, and hence, there is such a path from v to any node of G_t. Since the node v was an arbitrary clone, we have shown there exists a path between any two nodes in G_{t+1}.   □

In the case where $n_0 = 1$, then G_0 is K_1. Note that G_1 is $\overline{K_2}$, and G_2 is the disjoint union of two edges. In particular, G_1 and G_2 are not connected. The subsequent lemma will provide insight into how many iterations a disconnected graph requires before becoming connected.

Lemma 3. *For all $t \geq 0$, if G_t is a graph with $n_t \geq 4$, then G_{t+1} is connected.*

Proof. We proceed by a proof by contraposition. Suppose then that G_{t+1} is disconnected, and so there exists two nodes u, v in G_{t+1} such that there is no path between them.

Case 1: u, v are both in $V(G_t)$.

In this case, there is no set of size $\lfloor \frac{n_{t-1}}{2} \rfloor$ that contains both u and v, since otherwise, a clone in G_{t+1} would be adjacent to both u, v. At each time-step t, we add a clone for every subset of size $\lfloor \frac{n_t}{2} \rfloor$; hence, it must be the case that $\lfloor \frac{n_t}{2} \rfloor < 2$ which implies $n_t \leq 3$. This satisfies the negation of the predicate, and we have proved the result in this case.

Case 2: Exactly one of u or v is not in $V(G_t)$; without loss of generality, say $u \in V(G_{t+1}) \backslash V(G_t)$.

As u is a clone it has degree $\lfloor \frac{n_{t-1}}{2} \rfloor$, and so has a neighbor x in G_t, whenever $n_t \geq 2$. Thus, there is no path from x to v in G_t, and we apply Case 1 using these two nodes.

Case 3: Both u, v are in $V(G_{t+1}) \backslash V(G_t)$.

Since there are at least two clones it must be the case that $\alpha_t \geq 2$, and so $n_t \geq 2$. There then exists some neighbor x of u in G_t and some neighbor y of v in G_t. We then have that there is no path from x to y in G_t and we apply Case 1 to these two nodes. The proof follows.   □

Our next result proves the 2-connectivity of graphs generated by the half-model.

Lemma 4. *The graph G_t is 2-connected whenever $t \geq 4$, regardless of the input graph G_0.*

Proof. Using the recursive formula for the number of edges at time t in the proof of Theorem 1, for any graph G_0, we have at least four nodes after two time-steps. Using Lemma 3, we require at least one additional time-step to ensure connectivity. Thus, regardless of the input graph G_0, it is the case that G_t is connected for $t \geq 3$. We now claim that whenever a graph G_t is connected, G_{t+1} will be 2-connected.

Claim: If G_t is connected and $n_t \geq 4$, then G_{t+1} is 2-connected.

If G_t is 2-connected, then we are done since every node in the set $V(G_{t+1}) \backslash V(G_t)$ has at least one neighbor in $V(G_t)$, and we may use the same two paths between those neighbors to find 2-connectivity. Suppose G_t is at most 1-connected and thus let u be a cut-node of G_t. Consider two nodes in G_t, say a, b, that have a shortest path through u. In G_{t+1}, there is some clone z that is adjacent to both a, b. Therefore, we have two paths from a to b, and the proofs of the claim and theorem follow.   □

Our main result on the diameter of half-model graphs is the following.

Theorem 3. *Suppose that G_0 has order at least 4. In the half-model, the diameter of G_t for $t \geq 5$, is at most three.*

Proof. We consider the distance between two non-adjacent nodes $x, y \in V(G_t)$ in three cases.

Case 1: $x, y \in V(G_{t-1})$.

There exists some set $S \subseteq V(G_{t-1})$ of cardinality $\lfloor \frac{n_{t-1}}{2} \rfloor$ containing both x and y. Thus, the dominating node for this set S, v_S is adjacent to both x and y so their distance is 2.

Case 2: $x \in V(G_{t-1})$ and $y \notin V(G_{t-1})$.

There exists a node $z \in N_{G_t}(y)$. There is some set $S \subseteq V(G_{t-1})$ so that $x, z \in S$. The node v_S that dominates S in G_t is adjacent to both x and z, so we have the path yzv_Sx. Hence, the distance between x and y is at most 3. The symmetric case where $y \in V(G_{t-1})$ and $x \notin V(G_{t-1})$ is analogous.

Case 3: $x, y \notin V(G_{t-1})$.

Since x, y are new nodes in time-step t, there must be two sets $S_x, S_y \subseteq V(G_{t-1})$, where x dominates S_x and y dominates S_y. If $S_x \cap S_y \neq \emptyset$, then there is some node of G_{t-1} adjacent to both x and y, so their distance is 2. Suppose now that $S_x \cap S_y = \emptyset$. Since $|S_x| = |S_y| = \lfloor \frac{n_{t-1}}{2} \rfloor$, it may be the case that there exists a node $z \notin S_x \cup S_y$.

Suppose first that there is no such node z. There must be some edge with one endpoint in S_x and the other in S_y, since otherwise, the graph would be disconnected, which contradicts Lemma 4. We call these two endpoints a and b. We then have a path $xaby$ and the distance between x and y is 3.

If there is such a node z, then since G_t is 2-connected by Lemma 4, z cannot be a cut-node. Therefore, there must be some edge with one endpoint in S_x and the other in S_y and the distance between x and y is 3.   □

3   Graph Parameters for the Half-Model

In this section, we discuss classical graph parameters for the half-model. We call $S \subseteq V$ a *dominating set* for G if for all $v \notin S$, there exists $w \in S$ such that $vw \in E$. The minimum cardinality of all dominating sets in G is denoted $\gamma(G)$, and is called the *domination number* of G. To colour a graph, we assign a colour from the set $\{c_0, c_1, \ldots, c_k\}$ to each vertex. A *proper colouring* is achieved when no two neighbouring vertices have the same colour. The *chromatic number* of a graph G, denoted by $\chi(G)$, is the minimum number of colours required to achieve a proper colouring of G. If G can be coloured using at most k colours, then we say that G is *k-colourable*. The maximum order of a clique in G is called the *clique number* of G, denoted by $\omega(G)$. The maximum order of an independent set in G is called the *independence number* of G, denoted by $\alpha(G)$.

For further background on these parameters, the reader is directed to [13]. We begin by considering the independence and clique number.

Theorem 4. *The independence number of G_t is α_{t-1} and for the clique number we have*

$$\omega(G_t) \geq \min\left(\left\lfloor \frac{n_{t-1}}{2} \right\rfloor + 1, \omega(G_0) + t\right).$$

Proof. At each time-step t, all the cloned nodes form an independent set. The set of new nodes has order $\alpha_{t-1} \geq n_{t-1}$, so this set must be the largest independent set in G_t. Therefore, we derive that $\alpha(G_t) = \alpha_{t-1}$.

We next consider the clique number of G_t. At each time-step t, we add a dominating node to subsets of cardinality $\lfloor \frac{n_{t-1}}{2} \rfloor$ from G_{t-1}. If the largest clique K at time-step $t-1$ is contained in one such subset, then we have increased the order of K by 1. However, the maximum degree of new nodes is $\lfloor \frac{n_{t-1}}{2} \rfloor$. Hence, we cannot increase the size of the largest clique to be larger than $\lfloor \frac{n_{t-1}}{2} \rfloor + 1$. $\square$

We next give the chromatic number of the half-model.

Theorem 5. *For the half-model, we have that the chromatic number is given by*

$$\chi(G_t) = \min \left(\chi(G_0) + t, \left\lfloor \frac{n_{t-1}}{2} \right\rfloor + 1 \right).$$

Proof. Suppose that G_t is properly colored. Consider a *rainbow* subset of nodes; that is, a set of nodes that requires each distinct color in the graph. Let the cardinality of this set be $r \geq 1$. When $r \leq \lfloor \frac{n_{t-1}}{2} \rfloor$, any new clone that is added contains this set in its neighbors will need a new color. When $r > \lfloor \frac{n_{t-1}}{2} \rfloor$, any new clone that is added will have a neighbor set smaller than the cardinality of the colors, which implies there will always be an available color. $\square$

We finish by proving a result on the domination number of graphs generated by the half-model.

Theorem 6. *The domination number of G_t is*

$$\gamma(G_t) = \left\lceil \frac{n_{t-1}}{2} \right\rceil + 1.$$

Proof. We will first establish the upper bound

$$\gamma(G_t) \leq \left\lceil \frac{n_{t-1}}{2} \right\rceil + 1.$$

Consider a set S of $\lfloor \frac{n_{t-1}}{2} \rfloor$ non-clone nodes in G_{t-1}. The node x_S dominates S. The complement T of S in $V(G_{t-1})$ has cardinality $\lceil \frac{n_{t-1}}{2} \rceil$. Hence, $T \cup \{x_S\}$ is the desired dominating set.

For the lower bound, we must show that $\gamma(G_t) > \lceil \frac{n_{t-1}}{2} \rceil$. For a contradiction, suppose that some set of $\lceil \frac{n_{t-1}}{2} \rceil$-many nodes, say X, dominates G_t. Suppose first that X consists of non-clones. Regardless of the choice of X, there will be some set of non-clones, call it T, of size $\lfloor \frac{n_{t-1}}{2} \rfloor$ such that $X \cap T = \emptyset$. Thus, x_T is not dominated, which is a contradiction.

Suppose that X contains at least one clone. There is a least one clone z not adjacent to $X \cap V(G_{t-1})$, since $|X \cap V(G_{t-1})| < \lceil \frac{n_{t-1}}{2} \rceil$. See Fig. 2. Any clone in X is not adjacent to z, since the clones form an independent set. Therefore, z is not dominated by X, which gives a contradiction. $\square$

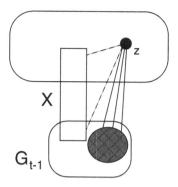

Fig. 2. The node z is not adjacent to X.

4 Conclusion and Further Directions

We introduced the Iterated Global Model (IGM) for complex networks. The IGM adds new nodes joined to $\lfloor \frac{1}{k} n_t \rfloor$, where n_t is the number of nodes at time t. Our focus was the case $k = 2$, and we proved that graphs generated by the half-model exhibit densification, low distances, and bad spectral expansion as found in real-world, complex networks. We investigated various classical graph parameters for this model, including the clique, chromatic, and domination numbers.

Several open problems remain concerning properties of graphs generated by the half-model. Graph limits consider dense sequences of graphs and analyze their properties based on their homomorphism densities; see [11]. Since the half-model generates dense sequences of graphs, it would be interesting to explore their graph limits. In the full version, we will consider the clustering coefficient of the half-model, analyze its subgraph counts, and degree distribution. Another interesting direction would be to generalize our results to integers $k > 2$.

References

1. Bonato, A.: A Course on the Web Graph. Graduate Studies Series in Mathematics. American Mathematical Society, Providence (2008)
2. Bonato, A., Chuangpishit, H., English, S., Kay, B., Meger, E.: The iterated local model for social networks (2020). Accepted to Discrete Applied Mathematics
3. Bonato, A., Infeld, E., Pokhrel, H., Prałat, P.: Common adversaries form alliances: modelling complex networks via anti-transitivity. In: Bonato, A., Chung Graham, F., Prałat, P. (eds.) WAW 2017. LNCS, vol. 10519, pp. 16–26. Springer, Cham (2017). https://doi.org/10.1007/978-3-319-67810-8_2
4. Bonato, A., Hadi, N., Horn, P., Prałat, P., Wang, C.: Models of on-line social networks. Internet Math. **6**, 285–313 (2011)
5. Bonato, A., Tian, A.: Complex networks and social networks. In: Kranakis, E. (ed.) Advances in Network Analysis and its Applications. Mathematics in Industry series, pp. 269–286. Springer, Heidelberg (2011). https://doi.org/10.1007/978-3-642-30904-5_12

6. Chung, F.R.K.: Spectral Graph Theory. American Mathematical Society, Providence (1997)
7. Chung, F.R.K., Lu, L.: Complex Graphs and Networks. American Mathematical Society, Providence (2006)
8. Easley, D., Kleinberg, J.: Networks, Crowds, and Markets Reasoning About a Highly Connected World. Cambridge University Press, Cambridge (2010)
9. Estrada, E.: Spectral scaling and good expansion properties in complex networks. Europhys. Lett. **73**, 649–655 (2006)
10. Leskovec, J., Kleinberg, J., Faloutsos, C.: Graphs over time: densification laws, shrinking diameters and possible explanations. In: Proceedings of the 13th ACM SIGKDD International Conference on Knowledge Discovery and Data Mining (2005)
11. Lovász, L.: Large Networks and Graph Limits. American Mathematical Society, Providence (2012)
12. Spencer, J., Florescu, L.: Asymptopia. American Mathematical Society, Providence (2014)
13. West, D.B.: Introduction to Graph Theory, 2nd edn. Prentice Hall, Upper Saddle River (2001)

A Robust Method for Statistical Testing of Empirical Power-Law Distributions

Davide Garbarino[1]([⊠]), Veronica Tozzo[1], Andrea Vian[2], and Annalisa Barla[1]

[1] DIBRIS, Universitá di Genova, Genova, Italy
davide.garbarino@edu.unige.it, veronica.tozzo@dibris.unige.it,
annalisa.barla@unige.it
[2] DAD, Universitá di Genova, Genova, Italy
andrea.vian@unige.it

Abstract. The World-Wide-Web is a complex system naturally represented by a directed network of documents (nodes) connected through hyperlinks (edges). In this work, we focus on one of the most relevant topological properties that characterize the network, *i.e.* being scale-free. A directed network is scale-free if its in-degree and out-degree distributions have an approximate and asymptotic power-law behavior. If we consider the Web as a whole, it presents empirical evidence of such property. On the other hand, when we restrict the study of the degree distributions to specific sub-categories of websites, there is no longer strong evidence for it. For this reason, many works questioned the almost universal ubiquity of the scale-free property. Moreover, existing statistical methods to test whether an empirical degree distribution follows a power law suffer from large sample sizes and/or noisy data.

In this paper, we propose an extension of a state-of-the-art method that overcomes such problems by applying a Monte Carlo sub-sampling procedure on the graphs. We show on synthetic experiments that even small variations of true power-law distributed data causes the state-of-the-art method to reject the hypothesis, while the proposed method is more sound and stable under such variations.

Lastly, we perform a study on 3 websites showing that indeed, depending on their category, some accept and some refuse the hypothesis of being power-law. We argue that our method could be used to better characterize topological properties deriving from different generative principles: central or peripheral.

Keywords: Power-law distribution · Monte Carlo · Statistical test · World-Wide-Web · Network Analytics

1 Introduction

The World-Wide-Web (WWW) encodes associative links among a large amount of pages. Its structure has grown without any central control, thus make it approximable to the result of a random process, where pages link to each other following local probabilistic rules.

© Springer Nature Switzerland AG 2020
B. Kamiński et al. (Eds.): WAW 2020, LNCS 12091, pp. 145–157, 2020.
https://doi.org/10.1007/978-3-030-48478-1_11

Such probabilistic rules are defined through statistical properties of Web graph features. In particular, several investigations show that the WWW is scale-free [1,5,10] *i.e.*, both the distributions of incoming and outgoing links are well-approximated by a discrete power law [21]. This can be traced to the fact that the vast majority of documents in the Web have relatively few outgoing and incoming links, but few pages still have enormous number of links that skew the mean of the empirical distribution far above the median.

Nonetheless, when analyzing specific portions of the Web, *i.e.* websites, the scale-free property seems to be less evident especially for specific categories (*e.g.* university homepages) [22,24]. Note that, differently from what is commonly done in literature [22], we consider websites as closed sub-systems of the Web whose temporal evolution is independent of the system they evolved into.

In this work, we are interested in developing a method able to assess if data from empirical observations follow a power-law. Indeed, testing power laws on empirical data is usually hard due to the large fluctuations that are present in the tail of the distribution.

One of the most commonly used statistical test is the Kolmogorov-Smirnov [11]. This method focuses on the center of the distribution, making it not suitable for testing heavy-tailed distributions. In [11] the authors make strong use of this test by performing a bootstrap procedure that is optimal in small sample size regimes. Indeed, as the sample size grows, the power of the statistical test increases, thus leading to higher rate of rejections of the null hypothesis. Moreover, even in presence of small sample sizes, adding a low amount of noise may cause the test to reject.

As in real-world, noisy or large samples are the common scenario, here, we propose an alternative testing pipeline that leverages on the Anderson Darling test [3] and Monte Carlo sub-sampling. Our pipeline is able to cope with the power of the test problem by reducing the sample size while maintaining the original degree distribution behavior.

We show synthetic experiments in which the state-of-the-art method fails under small variations or large sample sizes of input data. In all these cases, our method is proved to be more stable under variations and it can be shown that provides results with a better confidence. Lastly, we present case studies on 3 websites representative of different generative processes. These case studies present interesting results showing that indeed, closed sub-portion of the Web do not necessarily follow a power-law distribution. And, they seem to point in the direction that the more the generative process is centralized the less the degree distribution can be associated to a power law decay.

Outline. The remainder of the paper is organized as follows: Sect. 2 presents the state-of-the art algorithm for testing empirical power-law distribution; in Sect. 3 we present the limitations of such method with the related synthetic examples; in Sect. 4 we present our adaptation based on Monte Carlo sub-sampling to overcome the issue of power in empirical data; in Sect. 5 we present a large variety of experiments showing how our method is more stable and the case studies; lastly, we conclude with Sect. 6 with some discussion on the obtained results and future research directions.

2   Discrete Power-Law Distribution: Definition, Fit and Statistical Test

The discrete power-law distribution is defined as

$$\mathbb{P}(d_v = x) \approx \frac{1}{\xi(x_{min}, \alpha)} x^{-\alpha}, \tag{1}$$

where d_v is the random variable representing the degree of a node v, x_{min} is a fixed lower bound on the values x, α is a *scaling parameter*, and $\xi(x_{min}, \alpha) = \sum_{x=x_{min}}^{\infty} x^{-\alpha}$ is the Hurwitz-zeta function [15].

The parameter x_{min} is particularly important, as often the degree distribution of a network follows a power law only for degrees x greater than a lower bound. A network is said to be scale-free if the tail of its in-degree and out-degree distributions obeys to a discrete power law decay. In practice, this entails that we have a non-null probability to observe nodes with a degree much greater than average (hubs).

2.1   Maximum Likelihood Estimation

The parameters x_{min} and α of an empirical power-law distribution need to be estimated from data. Given as input a vector $\mathbf{x} \in \mathbb{N}^n$ representing the degrees of n nodes of a graph, we need to perform two different procedures to estimate these two parameters, as described by the pseudo-code in Algorithm 1.

Estimate of x_{min}. First, we pick $\hat{x}$ as the value that minimizes the difference between the empirical degree distribution and the fitted power-law model where $x_{min} = \hat{x}$ [11,12].

In order to minimize such difference, we need to select a suitable distance. One of the most common is the Kolmogorov-Smirnov (KS) statistic, which is defined as the supremum norm of the difference between two distribution functions (CDFs) of the empirical data and the best-fit model [18]. Although the KS statistic is widely used, it presents some drawbacks in the detection of heavy-tailed distributions since, being based on the CDF, it mainly penalizes fluctuations in the center of the empirical distribution. A more reliable distance for the comparison of heavy-tailed distributions is the Anderson-Darling (AD) statistic as it puts more importance to the extreme values of the CDFs [3]. For this reason, we will recur to this statistic in the rest of the paper. The AD distance is defined as

$$A^2(\mathbf{x}, F_{x_{min}=x}) = -n - \sum_{i=1}^{n} \frac{2i-1}{n} \left[\ln F_{x_{min}=x}(x_i) + \ln(1 - F_{x_{min}=x}(x_{n+1-i})) \right],$$

Algorithm 1. Power-law fitting

1: **Input**: degrees vector of length n
2: distances $= [\,]$
3: **for** $x \in \{\min(\text{degrees}), \ldots, \max(\text{degrees})\}$ **do**
4:     **if** len(unique(degrees) $> x$) <25 **then**
5:         **break**
6:     $\alpha \leftarrow$ power_law_fit(degrees, $x_{min} = x$)
7:     $d \leftarrow$ Anderson-Darling(degrees, x, α)
8:     distances.append(d)
9: $\hat{x} \leftarrow \underset{x}{\text{argmin}}$ distances
10: $\hat{\alpha} \leftarrow$ power_law_fit(degrees, $x_{min} = \hat{x}$)
11: $\hat{d} \leftarrow$ Anderson-Darling(degrees, $\hat{x}, \hat{\alpha}$)
12: **return** $\hat{x}, \hat{\alpha}, \hat{d}$

---

where n is the sample size and $F_{x_{min}=x}$ is the power-law CDF.

Note that, if we select a $\hat{x} > x_{min}$, we are reducing the size of our training data, and our model will suffer from the statistical fluctuations in the tail of the empirical distribution. On the other hand, if $\hat{x} < x_{min}$, the maximum likelihood estimate of the scaling parameter $\hat{\alpha}$ may be severely biased.

Estimate of α. Given the lower bound x_{min}, we estimate the scaling parameter α by means of maximum likelihood, which provides consistent estimates in the limit of large sample sizes [13].

In the discrete case, a good approximation of the true scaling parameter can be reached mostly in the $x_{min} \geq 6$ regime [11]. And it can be computed as:

$$\hat{\alpha} \approx 1 + n \left[\sum_{i=1}^{n} \ln \frac{x_i}{x_{min} - \frac{1}{2}} \right]^{-1}.$$

2.2 Goodness-of-Fit Test

Once $\hat{\alpha}$ and $\hat{x}$ have been estimated, we need to assess if observed data are plausibly sampled from the related power-law distribution. To such extent, we perform a goodness-of-fit (GoF) test procedure [19].

A goodness-of-fit test measures how well a statistical model fits into a set of observations. Given the statistical model under testing, a GoF makes use of a statistic that evaluates the discrepancy between the observed values and the expected value of the model. By definition, a statistic is a function which does not depend on the parameters of the model. The output of the GoF procedure is a p-value corresponding to the probability that the statistic is greater than its realization on the observed data.

Note that, since we estimate the model parameters from data we do not know the distribution of the statistic. Thus, we perform a semi-parametric bootstrap approach to estimate such distribution empirically [11,25].

Algorithm 2. Power-law testing

1: **Input**: degrees vector of length $n, \hat{x}, \hat{\alpha}, \hat{d}$
2: distances $= [\,]$
3: **for** $i = 1, \ldots, M$ **do**
4:     $n_{tail} = \text{count}(\text{degrees} > \hat{x})$
5:     **for** $j = 1, \ldots, n$ **do**:
6:         $b \leftarrow \text{bernoulli\_sample}(n_{tail}/n)$
7:         **if** b is 1 **then**
8:             $s_i[k] = \text{power\_law\_sample}(\hat{x}, \hat{\alpha})$
9:         **else**
10:             $s_i[j] \leftarrow \text{uniform\_sample}(\text{degrees} < \hat{x})$
11:     $\alpha_i, x_i \leftarrow \text{power\_law\_fit}(s)$
12:     $d \leftarrow \text{Anderson-Darling}(s, x_i, \alpha_i)$
13:     distances.append(d)
14: p-value $= \text{count}(\text{distances} > \hat{d})/M$
15: **return** p-value

In particular, we fixed as statistic the Anderson-Darling distance and we perform a procedure described in Algorithm 2. Given n samples, we indicate with n_{tail} the amount of samples that are greater than $\hat{x}$. Bootstrap is then performed by simulating n_{tail} examples from a power law with parameters $\hat{\alpha}$ and $\hat{x}$, and for the remaining sample size $n - n_{tail}$ we sample degrees from the empirical data that are smaller than $\hat{x}$. We repeat this procedure M times. The value of M depends on the desired significance of the p-value. Typically, if we want a p-value that approximates its true value with an error smaller than ϵ, then $M = \frac{1}{4\epsilon^2}$.

Given the M simulated data sets, we fit to each of them its own power-law model and compute the AD distance. This provides the empirical distribution of the AD statistic that we use to compute the associated p-value, defined as the fraction of synthetic distances larger than the observed one.

If p is large (relatively to a fixed significance level, *e.g.* 0.1), we cannot reject the null hypothesis. Then, possibly, the difference between the empirical and theoretical distributions may be attributed to statistical fluctuations. Differently, if p is smaller than the significance level, we say that the empirical data are not power law.

3   Problems of Goodness-of-Fit on Empirical Data

Testing whether empirical data are power-law distributed is a hard task. This is due to the following reasons: a) the probability of rejecting the null hypothesis grows with sample size; and, as a consequence b) the procedure is too sensitive to even minimal amount of noise. Little attention has been put on these issues, but we argue that they are crucial as they heavily affect the final response of the statistical test.

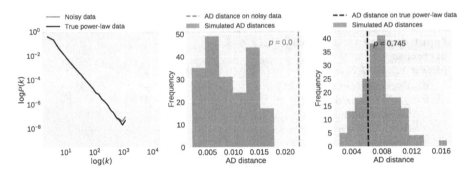

Fig. 1. On the left, the empirical probability density functions of true power-law data (black line) and noisy power-law data (pink). On the right, the Anderson-Darling test on both samples. Little variations from an exact power-law sample lead to reject the null hypothesis. (Color figure online)

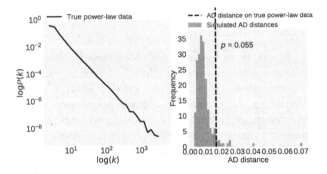

Fig. 2. On the left, the empirical probability density function of true power law data. On the right, the Anderson-Darling test. Large sample size (5×10^5) leads to reject the null hypothesis.

In particular, both problems can be addressed by considering the *power of the test*, which, fixed a significance level, is defined as the probability of correctly rejecting the null hypothesis. Such probability increases accordingly to the sample size, hence, when the number of nodes n is large, we tend to reject the null hypothesis even in cases of true power-law distributed data (as the power of the test is very close to 1). Indeed, by performing bootstrap, we simulate nearly exact power-law samples, which induce the Anderson-Darling test to be very sensitive to even minimal fluctuations in the observed distribution.

In Fig. 1 and 2, we show two synthetic experiments where such test fails, in particular:

(a) we generated $n = 10^5$ samples from a discrete power-law distribution with parameters $x_{min} = 7$ and $\alpha = 2.7$. We perturbed the data by adding one occurrence to the last 13° in the extreme tail (see Fig. 1 left panel for the true and perturbed data);

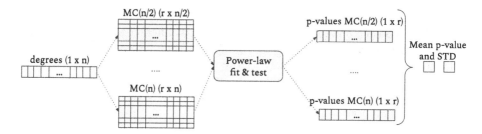

Fig. 3. Schematic representation of the proposed pipeline.

(b) we generated $n = 5 \times 10^5$ samples from a discrete power-law distribution with parameters $x_{min} = 2$ and $\alpha = 2.7$.

We applied the procedure in Sect. 2 on both datasets, with $M = 200$ and significance level set to 0.1. Results are shown on the right side of Fig. 1 and 2. In Fig. 1, the empirical probability density functions of the two samples are indistinguishable from each other except in the extreme tail, where little divergences can be traced. Thus, it becomes evident that for large sample sizes the test is very sensitive even to little fluctuations in the observed sample. Also, with example (b) we show that even perfect power-law samples induce the test to fail when the sample size is too large (Fig. 2).

Both examples show that the high power of the Anderson-Darling test in large sample size regimes constitutes a drawback of the previously introduced method [11]. Since it is never the case that an observed degree distribution is exactly drawn from a discrete power law, we propose a variation of the method in Sect. 2 that aims at testing the goodness of fit of heavy tail distributions.

4   Monte Carlo Approach

Our proposal is based on the idea of performing iterative Monte Carlo (MC) sub-samplings of different length on the original degree sequence. We argue that with this sub-sampling scheme we can reduce the sample size without modifying the trend of the original degree distribution and possibly obtain a more reliable test.

The global scheme of the procedure is provided in Fig. 3. In particular, we define a set of lengths, $\{l_1, \ldots, l_{max}\}$, for each length we perform r corresponding MC samplings. For each sample, we fit a power-law distribution and assess its plausibility exploiting Algorithm 1 and Algorithm 2 and, thus, obtaining a sequence of p-values of the Anderson-Darling test of length r. We consider, as final output of the procedure, the mean of all p-values sequences for all different lengths and the related standard deviation.

To the best of our knowledge, it is not usual to exploit MC sub-sampling to test for power-law decay in the degree distribution. In fact, performing MC does not allow to exactly estimate the parameters of the power-law distribution, indeed, to each sub-sample may correspond a different set of parameters. Nonetheless, we do not use MC as a fitting method but rather to say if a network is plausible to asymptotically satisfying the scale-free property. We argue that using MC as a way to obtain suitable sub-samples of smaller sample size would provide better understanding of the degree sequence behavior while overcoming the drawbacks induced by large sample sizes.

4.1   Instantiation of Parameters

In order to apply the Monte Carlo approach we need to fix different values, specifically l_1, l_{max}, r and the significance level.

The problem of selecting adequate lengths for the MC sub-samples is not trivial. On the one hand, a too small sub-sample would lead to very different degree sequences due to the large fluctuations present in the original network, while, on the other hand, lengths close to the original degree sequence would lead to higher rates of rejection of the power-law hypothesis. Then, we arbitrarily decided to set l_1 at $n/2$ which is half the length of the observed data. As for l_{max}, we fix it to n as in case of true power-law samples we want to being able to obtain a high p-value, while in case of noisy data, considering one length equal to the original size does not particularly affect the resulting mean p-value.

The value of r affects the robustness of the final result, the more repetitions the better approximation of the true p-value. Nonetheless, its value depends on constraints deriving from computational power. Thus, we leave the definition of such value to the user.

We fixed the significance level at 0.1 for the rejection of the null hypothesis. This is a conservative choice implying that the power law hypothesis is ruled out if there is a probability of 1 in 10 or less that data sampled from the true model agree with the model as the empirical data.

Lastly, we fixed the maximal possible x_{min} to be least 25 observations less than the maximal observed degree. This is due to limit the chances of fitting a power-law distribution on too few observations.

5   Experimental Results

In order to evaluate the performance of the proposed pipeline, we perform four experiments and compare the results with the state-of-the-art method. In the rest of the narration we will refer to the state-of-the-art method as Bootstrap and to our method as Monte Carlo + Bootstrap.

All the simulations are performed in Python. We used the package **powerlaw** [2] for fitting power-law distributions to empirical data and compute the AD distances. We provide all the notebooks used for the experiments of this paper in a GitHub repository[1]. For all experiments, we fixed 30 lengths of Monte Carlo re-sampling in the interval $[\frac{n}{2}, n]$ and for each of this length we get $r = 10$ re-samplings.

---

[1] https://github.com/DaviGarba/netanalytics.

Table 1. Results to assess the goodness of the proposed testing pipeline in cases of scale-free graphs (Barabasi-Albert) or not (Erdős-Renyi), in terms of mean p-value and standard deviation on 10 repetitions of the test for different sample sizes.

Test type	Erdős-Renyi		Barabasi-Albert		
	75000	150000	75000	150000	300000
Bootstrap	0.00 ± 0.00	0.00 ± 0.00	0.85 ± 0.24	0.75 ± 0.27	0.78 ± 0.18
MC + Boostrap	0.00 ± 0.00	0.00 ± 0.00	0.85 ± 0.06	0.69 ± 0.16	0.71 ± 0.15

5.1 Validation of the Proposed Method on Different Graph Models

In the first experiment we aim at verifying if Monte Carlo + Bootstrap is comparable to just Bootstrap when considering two cases at varying sample sizes:

1. Erdős-Renyi models of size $\{75 \times 10^3, 15 \times 10^4\}$, we expect both methods to refuse the null hypothesis as the degree distribution of this model is known to follow a binomial distribution [14]. Thus, we use this as base test to assess the probability of correctly rejecting the power-law hypothesis.
2. Barabasi-Albert models of size $\{75 \times 10^3, 15 \times 10^4, 3 \times 10^5\}$, we expect both methods to have high p-values as the degree distribution follows a power law [6]. We use this experiment to provide proof of the soundness of the method in presence of true power-law data.

Each experiment listed above is repeated 10 times to estimate the mean and standard deviation of p-values. Results are reported in Table 1 where we observe that our approach (Monte Carlo + Bootstrap) always reject the null hypothesis in the Erdős-Renyi case as the Bootstrap method, while in the Barabasi-Albert case we always provide p-values with a smaller variance.

5.2 Robustness to Noise

We now want to assess that our method is indeed more robust under increasing noise in the input empirical distribution. We simulated from a discrete power law with parameters $\alpha = 2.3$ and $x_{min} = 1$, a sample of size $n = 10^5$. For different levels of noise in the set $\bar{n} \in \{10, 40, 70, 100\}$, we perturbed the power law observation by adding $\bar{n}$ values uniformly sampled from the original observation.

Figure 4 shows that the proposed methods is in mean always better than the simple bootstrap approach while also providing a smaller variance. Also, it never reject the null-hypothesis in cases in which the noise is small while sometimes it rejects it in presence of high amount of noise (100 added observations). Differently from the Bootstrap approach that, depending on the simulated sample, sometimes rejects it even in presence of zero noise.

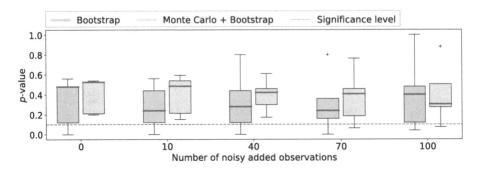

Fig. 4. Results in terms of p-values for the two testing pipelines as the input data present an increasing level of noisy observations.

5.3   Benchmark on University of Notre Dame Website

We exploit a widely studied example of empirical data that is assumed to follow a power-law distribution [1,4,20], *i.e.* the web graph of the University of Notre Dame website. This graph, in 1999, has been studied in order to obtain information regarding the topology of the Web. In [1], the authors found that the in-degree and out-degree distributions of the graph underlying the hyperlink structure of the domain nd.edu were well approximated by power-law distributions with scaling parameters 2.7 and 2.1 respectively. We downloaded the hyperlink graph from http://snap.stanford.edu/ [16]; the crawl consists of 325729 documents and 1497134 links. We tested the Monte Carlo + Bootstrap approach against the Bootstrap approach as the empirical data are noisy and we want to provide further validation of our testing procedure on the in-degree distribution of the network.

We performed $r = 5$ MC re-samplings for different sizes equally spaced in the interval [162864, 325729]. Monte Carlo + Bootstrap results in a mean p-value of 0.15, meaning that there is no strong evidence against the power-law hypothesis for the in-degree distribution. Differently, when applying the Bootstrap method we observed a p-value equal to 0.00, which would lead us to reject the null hypothesis.

As in literature many have argued the power-law nature of this graph, this allows us to conclude that our testing procedure is more robust and thus can be applied on real-world data with higher reliability.

5.4   Websites Analysis

We now want to exploit our procedure in real scenarios to seek for evidence of differences in the degree distributions deriving from different generative processes. We considered three different websites that we deemed representative of different strategies of content creation: e-commerce, academic and free encyclopedia. The first category is typically characterized by a strong central control in the design and evolution of the information architecture and content generation. Conversely, the last category is completely user-guided and its evolution is, thus, likely to be random. We argue that the academic category, as well as other website of complex institutions, should be a trade-off between the two, as usually many contributors have access to writing and adding content with a mild central control (Fig. 5).

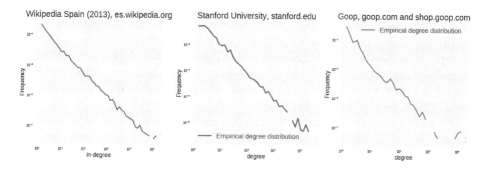

Fig. 5. Log-log plots of the empirical distributions of the considered case studies.

Table 2. Analyzed websites with the related information about number of nodes, number of edges and category.

Name	Url	Website type	No. nodes	No. edges	p-value
Goop	goop.com	E-commerce	100.482	731.259	0.00 ± 0.00
Stanford	stanford.edu	Academic	281.903	2.312.497	0.01 ± 0.00
Wikipedia (ES)	es.wikipedia.org	Encyclopedia	972.933	23.041.488	0.74 ± 0.21

We consider the following websites:

1. Goop, the website of a wellness and lifestyle company; we crawled the entire website using the open source framework Scrapy[2], during the crawl we restricted to the domains **goop.com** and **shop.goop.com**;
2. Stanford, the website of Stanford University. We downloaded a crawl performed in 2002 available at http://snap.stanford.edu/;
3. Wikipedia (ES), the website of the free spanish encyclopedia. We downloaded a crawl of 2013 at http://law.di.unimi.it/index.php [8,9].

Table 2 describes the characteristics of the three considered websites, in terms of category, number of nodes and number of edges. Table 2 also reports the mean p-values obtained with Monte Carlo + Bootstrap on the in-degree distributions. Results seems to validate our hypothesis about an inverse correlation between the centrality of the content generative process and the scale-free property.

---

[2] https://scrapy.org/.

6   Discussion

In this paper, we proposed a method for hypothesis testing of power-law distributions in empirical data that overcomes issues related to the power of the test. In particular, our method mediates the effect of possibly noisy data through Monte-Carlo sub-samplings of the empirical distribution. We verified that the proposed method retains the ability of assessing if observations are indeed plausibly sampled from a power law, under different sample sizes and level of noise. Indeed, the method is more reliable than the state-of-the art on synthetic data. To further assess the reliability of our approach we also provide a real-world example, specifically the University of Notre Dame website, which is a well studied dataset and it is considered to be scale-free. Our method does indeed provide a p-value higher than the significance level, differently from the state-of-the-art method that rejects the null hypothesis.

This allowed us to use our method to test different websites corresponding to different content generative processes. From a first insights, we observed that different content generation strategies may induce a different connectivity structure of the hyperlink graph.

For future research we intend to increase the number of real networks studied and consider current websites related to different generative processes to provide a more comprehensive understanding of specific sub-categories of the Web.

Future research directions may also involve the use of random walks instead of Monte Carlo as a sub-sampling technique on graphs [7,17] and the comparison with other estimators of power laws in empirical data [23].

To conclude, our pipeline is an attempt to perform statistical testing while considering its limits both theoretical and due to noisiness of data. We argue that this is fundamental to reliably test assumptions on real-world examples.

References

1. Albert, R., Jeong, H., Barabási, A.L.: Diameter of the world-wide web. Nature **401**(6749), 130–131 (1999)
2. Alstott, J., Bullmore, D.P.: Powerlaw: a Python package for analysis of heavy-tailed distributions. PLoS ONE **9**, 1 (2014)
3. Anderson, T.W., Darling, D.A.: A test of goodness of fit. J. Am. Stat. Assoc. **49**(268), 765–769 (1954)
4. Barabási, A.L., Albert, R.: Emergence of scaling in random networks. Science **286**(5439), 509–512 (1999)
5. Barabási, A.L., Albert, R., Jeong, H.: Scale-free characteristics of random networks: the topology of the world-wide web. Physica Stat. Mech. Appl. **281**(1–4), 69–77 (2000)
6. Barabási, A.L., et al.: Network Science. Cambridge University Press, Cambridge (2016)
7. Basirian, S., Jung, A.: Random walk sampling for big data over networks. In: 2017 International Conference on Sampling Theory and Applications (SampTA), pp. 427–431. IEEE (2017)

8. Boldi, P., Rosa, M., Santini, M., Vigna, S.: Layered label propagation: a multireso-
   lution coordinate-free ordering for compressing social networks. In: Srinivasan, S.,
   Ramamritham, K., Kumar, A., Ravindra, M.P., Bertino, E., Kumar, R. (eds.) Proceedings of the 20th International Conference on World Wide Web, pp. 587–596.
   ACM Press (2011)
9. Boldi, P., Vigna, S.: The WebGraph framework I: compression techniques. In:
   Proceedings of the Thirteenth International World Wide Web Conference (WWW
   2004), pp. 595–601. ACM Press, Manhattan (2004)
10. Broder, A., et al.: Graph structure in the web. Comput. Netw. **33**(1–6), 309–320
    (2000)
11. Clauset, A., Shalizi, C.R., Newman, M.E.: Power-law distributions in empirical
    data. SIAM Rev. **51**(4), 661–703 (2009)
12. Clauset, A., Young, M., Gleditsch, K.S.: On the frequency of severe terrorist events.
    J. Conflict Resolut. **51**(1), 58–87 (2007)
13. Daniels, H.: The asymptotic efficiency of a maximum likelihood estimator. In:
    Fourth Berkeley Symposium on Mathematical Statistics and Probability, vol. 1,
    pp. 151–163. University of California Press, Berkeley (1961)
14. Erdös, P., et al.: On random graphs
15. Hardy, M.: Pareto's law. Math. Intell. **32**(3), 38–43 (2010)
16. Leskovec, J., Sosič, R.: SNAP: a general-purpose network analysis and graph-
    mining library. ACM Trans. Intell. Syst. Technol. (TIST) **8**(1), 1 (2016)
17. Lovász, L., et al.: Random walks on graphs: a survey. Combinatorics **2**(1), 1–46
    (1993). Paul erdos is eighty
18. Massey Jr., F.J.: The Kolmogorov-Smirnov test for goodness of fit. J. Am. Stat.
    Assoc. **46**(253), 68–78 (1951)
19. Maydeu-Olivares, A., Garcia-Forero, C.: Goodness-of-fit testing. Int. Encycl. Educ.
    7(1), 190–196 (2010)
20. Mossa, S., Barthélémy, M., Eugene Stanley, H., Nunes Amaral, L.A.: Trunca-
    tion of power law behavior in "scale-free" network models due to information
    filtering. Phys. Rev. Lett. **88**, 138701 (2002). https://link.aps.org/doi/10.1103/
    PhysRevLett.88.138701
21. Newman, M.E.: Power laws, Pareto distributions and Zipf's law. Contemp. Phys.
    46(5), 323–351 (2005)
22. Pennock, D.M., Flake, G.W., Lawrence, S., Glover, E.J., Giles, C.L.: Winners don't
    take all: characterizing the competition for links on the web. Proc. Natl. Acad. Sci.
    99(8), 5207–5211 (2002)
23. Resnick, S.I.: Heavy-Tail Phenomena: Probabilistic and Statistical Modeling.
    ORFE. Springer, Newyork (2007). https://doi.org/10.1007/978-0-387-45024-7
24. Stumpf, M.P., Porter, M.A.: Critical truths about power laws. Science **335**(6069),
    665–666 (2012)
25. Stute, W., Manteiga, W.G., Quindimil, M.P.: Bootstrap based goodness-of-fit-
    tests. Metrika **40**(1), 243–256 (1993)

Community Structures in Information Networks for a Discrete Agent Population

Peter Marbach[(✉)]

Department of Computer Science, University of Toronto, Toronto, Canada
marbach@cs.toronto.edu

Abstract. Using a game-theoretic framework, we characterize the community structure that emerges in a social (information) network. Our analysis generalizes the results in [1,2] that were obtained for the case of a continuous population model for the agents in the social network, to the case of a discrete agent population model. We note that a discrete agent set reflects more accurately real-life information networks, and are needed in order to get additional insights into the community structure, such as for example the connectivity (graph structure) within in a community, as well as information dissemination within a community.

Keywords: Social and information networks · Community structure

1 Introduction

The information network that we consider in this paper was proposed and studied in [1,2]. The work in [1,2] presents a model to study how agents form communities in information networks in order to efficiently share/exchange, and where agents in the network obtain a certain utility for joining a given community. Using a game-theoretic framework, the analysis in [1,2] characterizes the community structures that emerge under this model using the concept of a Nash equilibrium. An interesting aspect of the analysis and results in [1,2] is that the model is indeed is able to provide interesting insights into the microscopic structure of information communities. For example, the characterization of how content is being produced in the network, i.e. which content each agent produces, indeed matches what has been experimentally observed in real-life social networks (see discussion in Sect. 2).

The model in [1,2] is based on a continuous agent population model. This model simplifies the analysis, but it also has its limitations and drawbacks. For example, it does not lend itself readily to study and characterize the graph structure of how agents within a community connect (interact) with each other in order to exchange content. The reason for this is that the graph structure is typically assumes a discrete set of agents in order to describe the connectivity between agents. We address this issue in this paper by extending the mode in [1,2] to the case of discrete agent population model. Our analysis shows that the

© Springer Nature Switzerland AG 2020
B. Kamiński et al. (Eds.): WAW 2020, LNCS 12091, pp. 158–172, 2020.
https://doi.org/10.1007/978-3-030-48478-1_12

results in [1,2] essentially carry over to the discrete agent model. In particular, the results in [1,2] are recovered in the limit as the agent population becomes dense. Due to space constraints we present in this paper only the results of our analysis, the proofs of the results can be found in [3].

2   Related Work

There exists extensive work, both experimental and theoretical, on the macroscopic properties of social network graphs such as the small world phenomena, shrinking diameter, the dimension of the social network graph, and the power-law degree distribution (see for example [4,5]). The difference between this body of work and the model presented here is that these existing models a) do not explicitly model and analyze the community structure of the social network, and b) do not focus on deriving (characterizing) microscopic properties of a community such as the content that each agent in a given community produces.

Related to the analysis in this paper is the work on content forwarding and filtering in social networks presented in [6,7]. In [6], Zadeh, Goel and Munagala consider the problem of information diffusion in social networks under a broadcast model where content forwarded (posted) by a user is seen by all its neighbors (followers, friends) in the social graph. For this model, the paper [6] studies whether there exists a network structure and filtering strategy that leads to both high recall and high precision. The main result in [6] shows that this is indeed the case under suitable graph models such as for example Kronecker graphs. In [7], Hegde, Massoulie, and Viennot study the problem where users are interested in obtaining content on specific topics, and study whether there exists a graph structure and filtering strategy that allows users to obtain all the content they are interested in. Using a game-theoretic framework (flow games), the analysis in [7] shows that under suitable assumptions there exists a Nash equilibrium, and selfish dynamics converge to a Nash equilibrium. The main difference between the model and analysis in [6,7] and the approach in this paper is that model and analysis in [6,7] does not explicitly consider and model community structures, and the utility obtained by users under the models in [6,7] depends only on the content that agents receive, but not on the content agents produce.

There exists an interesting connection between the modeling assumption made in [6,7], and a result obtained in [1,2] and in this paper (Proposition 1, Sect. 4). Both papers [6,7] make the modeling assumption that users produce content only on a small subset of content that they are interested in receiving. In [6] this assumption is supported by experimental results obtained on Twitter data that shows that Twitter users indeed tend to produce content on a narrower set of topics than they consume. The results presented in [1,2] and in this paper provide a formal validation/explanation for this assumption as it shows that under the proposed model it is optimal for agents (users) to produce content on a small subset of the content type that they are interested in consuming. This result illustrates that the proposed model is able to capture and explain important microscopic properties of information networks and communities.

3 Model

In this section we introduce the mathematical model that we use for our analysis. For this we assume that there exists a population of agents who are interested in sharing content. Agents differ in the type of content they are interested in, as well as their ability to produce content. Agents can form/join communities in order to maximize their utility for obtaining and producing content. We use the concept of a Nash equilibrium to characterize the community structures that emerge under this model. In the following we define formally a) the space of content that is being produced and shared in the network, b) the agents' interest in content as well as their ability to produce content, c) the utilities that agents in obtain in a given community, and d) the Nash equilibrium structure that we use for our analysis.

3.1 Content Space and Agent Model

For our analysis we assume that we are given a set of agents that produce and consume content. Furthermore we assume that each content item that is being produced belongs to a particular content type. One might think of a content type as a topic, or an interest, that agents have. Furthermore we assume that there exists a "measure" that characterizes how closely related two different content types are. To model this situation we assume that the type of a content item is given by a point x in a metric space, and the closeness between two content types $x, x' \in \mathcal{M}$ is then given by the distance measure $d(x, x')$ of the metric space $\mathcal{M}$.

Agents that share (produce and consume content) might have different interests, as well as different abilities to produce content. To model this situation we associate with each agent that consumes content a center of interest $y \in \mathcal{M}$. The center of interest y of the agent is the content type (topic) that an agent is most interested in. The probability that an agent with center of interest y is interested in a content item of type x is given by

$$p(x|y) = f(d(x, y)), \qquad x, y \in \mathcal{M}, \tag{1}$$

where $d(x, y)$ is the distance between the center of interest y and the content type x, and $f : [0, \infty) \mapsto [0, 1]$ is a non-increasing function. As the function f is non-increasing, the agent is more interested in content that is close to its center of interest y.

Similarly given an agent that produces content, the center of interest y of the agent is the content type (topic) that an agent is most adapt at producing. The ability of the agent to produce content of type $x \in \mathcal{M}$ is then given by

$$q(x|y) = g(d(x, y)), \qquad x, y \in \mathcal{M}, \tag{2}$$

where $g : [0, \infty) \mapsto [0, 1]$ is a non-increasing function. The interpretation of this function is as follows. If an agent with center of interest equal to y produces a content item of type x, then this content item will be relevant to content type

x with probability $q(x|y)$ given by Eq. (1). As the function g is non-increasing, the agent is better at producing content that is close to its center of interest y.

In the following we identify each agent by its center of interest $y \in \mathcal{M}$, i.e. agent y is the agent with center of interest being equal to y. Let $\mathcal{A}^{(d)} \subset \mathcal{M}$ be the set of agents that consume content in the information network, and let be $\mathcal{A}^{(s)} \subset \mathcal{M}$ be the set of agents that produce content. In addition we assume that the sets $\mathcal{A}^{(d)}$ and $\mathcal{A}^{(s)}$ are given by discrete sets.

For simplicity, we assume for the reminder of the paper that the metric space $\mathcal{M}$ is a subspace of $\mathbb{R}^n$, $n \geq 1$.

3.2 Information Community and Community Structure

An information community $C = (C_s, C_d)$ is given by a set $C_d \subset \mathcal{A}^{(d)}$ of agents that consume content in the community C, and a set $C_s \subset \mathcal{A}^{(s)}$ of agents that produce content in the community C. Note that by definition the sets C_s and C_d are discrete sets. For each agent $y \in C_s$ that produces content in the community C, we define the function $\beta_C(x|y)$, $x \in \mathcal{M}$, which characterizes the rate with which agent y produces content of type x in the community C. Similarly, for each agent $y \in C_d$ that consumes content in the community C, we define the fraction of time $\alpha_C(y)$, $0 \leq \alpha_C(y) \leq 1$, that agent y spends consuming content in the community C.

Using these definitions we next characterize for a given community $C = (C_s, C_d)$ the utility rate for content consumption $U_C^{(d)}(y)$ of an agent $y \in C_d$ that consumes content in the community C, as well as the utility rate for content production $U_C^{(s)}(y)$ of an agent $y \in C_s$ that produces content in the community C. For this, we assume that agents pay a cost c, $c > 0$, for reading/consuming a content item, where c is a processing cost that reflects the effort/time required by an agent to read a content item (and decide whether it is of interest or not). If the content item is of interest, then the agent receives a reward equal to 1; otherwise the agent receives a reward equal to 0. The time-average utility rate ("reward minus cost") for content consumption $U_C^{(d)}(y)$ of agent $y \in C_d$ in community C is given by (see [1–3] for a detailed derivation)

$$U_C^{(d)}(y) = \alpha_C(y) \int_{\mathcal{M}} \left[\sum_{z \in C_s} \beta_C(x|z) \Big[q(x|z)p(x|y) - c \Big] \right] dx. \qquad (3)$$

Similarly, the time-average utility rate for content production $U_C^{(s)}(y)$ of agent y is given by

$$U_C^{(s)}(y) = \int_{\mathcal{M}} \beta_C(x|y) \left[\sum_{z \in C_d} \alpha_C(z) \Big[q(x|y)p(x|z) - c \Big] \right] dx. \qquad (4)$$

As discussed in [1–3], the utility rate for content production can be interpreted as the reputation, or "reputation score", of agent y in the community C, i.e.

it captures how beneficial the contributions of a content producer y are for the community C.

Having defined a community $C = (C_s, C_d)$ in an information network, we next define a community structure for an information network. A community structure $\mathcal{S}$ is then given by a triplet $(\mathcal{C}, \{\alpha_C(y)\}_{y \in \mathcal{A}^{(d)}}, \{\beta_C(\cdot|y)\}_{y \in \mathcal{A}^{(s)}})$ where $\mathcal{C}$ is a set of communities $C = (C_d, C_s)$ that exist in the information network, and

$$\alpha_C(y) = \{\alpha_C(y)\}_{C \in \mathcal{C}} \text{ and } \beta_C(y) = \{\beta_C(\cdot|y)\}_{C \in \mathcal{C}}$$

indicate the rates with which agents consume and produce allocate content in the different communities $C \in \mathcal{C}$.

In the following we assume that the total content consumption and production rates of each agent can not exceed a given threshold, and we have that

$$||\alpha_C(y)|| = \sum_{C \in \mathcal{C}} \alpha_C(y) \leq E_p, \qquad y \in \mathcal{A}^{(d)},$$

where $0 < E_p \leq 1$, and

$$\cdot \, ||\beta_C(y)|| = \sum_{C \in \mathcal{C}} ||\beta_C(\cdot|y)|| \leq E_q, \qquad y \in \mathcal{A}^{(s)},$$

where

$$||\beta_C(\cdot|y)|| = \int_{x \in \mathcal{M}} \beta_C(x|y) dx$$

and $0 < E_q$. Finally, we require that for a given community structure $\mathcal{S}$ that for each community $C = (C_d, C_s) \in \mathcal{C}$ we have that

$$\alpha_C(y) > 0, \quad y \in C_d, \text{ and } ||\beta_C(y)|| > 0, \quad y \in C_s.$$

3.3 ϵ−Equilibrium

Having defined a community structure in an information network, we next consider the situation where agents consume and produce content in the different communities in order to maximize their utility rates. For this situation we use a game-theoretic approach to characterize the community structures that emerge in an information network.

Given a community structure $\mathcal{S} = (\mathcal{C}, \{\alpha_C(y)\}_{y \in \mathcal{A}^{(d)}}, \{\beta_C(\cdot|y)\}_{y \in \mathcal{A}^{(s)}})$, let $U_{\mathcal{S}}^{(d)}(y)$, $y \in \mathcal{A}^{(d)}$, be the total utility rate for content consumption (over all communities) that agent y receives under this community structure. More precisely, let $U_{\mathcal{S}}^{(d)}(y)$, $y \in \mathcal{A}^{(d)}$, be given by

$$U_{\mathcal{S}}^{(d)}(y) = \sum_{C \in \mathcal{C}} U_C^{(d)}(y) = \sum_{C \in \mathcal{C}} \alpha_C(y) \int_{x \in \mathcal{M}} \left[\sum_{z \in C_s} \beta_C(x|z) \Big[q(x|z) p(x|y) - c \Big] \right] dx.$$

Similarly, let $U_{\mathcal{S}}^{(s)}(y)$, $y \in \mathcal{A}^{(s)}$, be the total utility rate for content production (over all communities) that agent y receives under this community structure. The utility rate $U_{\mathcal{S}}^{(s)}(y)$, $y \in \mathcal{A}^{(s)}$ is given by

$$U_{\mathcal{S}}^{(s)}(y) = \sum_{C \in \mathcal{C}} U_C^{(s)}(y) = \sum_{C \in \mathcal{C}} \int_{x \in \mathcal{M}} \beta_C(x|y) \left[\sum_{z \in C_d} \alpha_C(z) \Big[q(x|y)p(x|z) - c \Big] \right] dx.$$

Using these definitions, in the following we analyze the situation where an agent $y \in \mathcal{A}^{(d)}$ in a given a community structure $\mathcal{S}$ changes its rate allocation from $\alpha_C(y)$ to $\alpha_C{}'(y)$ given by

$$\alpha_C{}'(y) = \{\alpha_C{}'(y)\}_{C \in \mathcal{C}}$$

such that $\sum_{C \in \mathcal{C}} \alpha_C{}'(y) \leq E_p$. More precisely, let $U_{\mathcal{S}}^{(d)}(\alpha_C{}'(y)|y)$ be the utility that agent $y \in \mathcal{A}^{(d)}$ obtains under the new allocation $\alpha_C{}'(y)$ (while all other agents keep their rate allocation fixed) given by

$$U_{\mathcal{S}}^{(d)}(\alpha_C{}'(y)|y) = \sum_{C \in \mathcal{C}} \alpha_C{}'(y) \int_{x \in \mathcal{M}} \left[\sum_{z \in C_s} \beta_C(x|z) \Big[q(x|z)p(x|y) - c \Big] \right] dx.$$

Similarly, given a community structure $\mathcal{S}$ we analyze the situation where an agent $y \in \mathcal{A}^{(s)}$ changes its rate allocation $\beta_C(y)$ to $\beta_C{}'(y)$ given by

$$\beta_C{}'(y) = \{\beta_C{}'(\cdot|y)\}_{C \in \mathcal{C}},$$

such that $||\beta_C{}'(y)|| \leq E_q$. Let $U_{\mathcal{S}}^{(s)}(\beta_C{}'(y)|y)$ be the utility rate that agent y receives under the new allocation $\beta_C{}'(y)$ (while all other agents keep their rate allocation fixed) given by

$$U_{\mathcal{S}}^{(s)}(\beta_C{}'(y)|y) = \sum_{C \in \mathcal{C}} \int_{x \in \mathcal{M}} \beta_C{}'(x|y) \left[\sum_{z \in C_d} \alpha_C(y) \Big[q(x|y)p(x|z) - c \Big] \right] dx.$$

Ideally each agent wants to choose an allocations for consuming and producing content in order to maximize its own utility rates. Here we use a slightly weaker criteria where we assume that agents change their current allocations only if the new allocations provides an increase in their utility rate by a factor that is at least equal to ϵ, $\epsilon > 0$.

More formally, we call a community structure

$$\mathcal{S}^* = \left(\mathcal{C}, \{\alpha_C^*(y)\}_{y \in \mathcal{A}^{(d)}}, \{\beta_C^*(\cdot|y)\}_{y \in \mathcal{A}^{(s)}}\right)$$

a ϵ−equilibrium if

a) for all agents $y \in \mathcal{A}^{(d)}$ we have that

$$U_{\mathcal{S}^*}^{(d)}(\alpha_C(y)|y) - U_{\mathcal{S}^*}^{(d)}(y) < \epsilon,$$

where $\alpha_C(y) = \arg\max_{\alpha_C{}'(y):||\alpha_C{}'(y)|| \leq E_p} U_{\mathcal{S}^*}^{(d)}(\alpha_C{}'(y)|y),$

b) for all agents $y \in \mathcal{A}^{(s)}$ we have that

$$U_{\mathcal{S}^*}^{(s)}(\beta_{\mathcal{C}}(y)|y) - U_{\mathcal{S}^*}^{(s)}(y) < \epsilon,$$

where $\beta_{\mathcal{C}}(y) = \arg\max_{\beta_{\mathcal{C}'}(y):||\beta_{\mathcal{C}'}(y)|| \leq E_q} U_{\mathcal{S}^*}^{(s)}(\beta_{\mathcal{C}'}(y)|y)$.

In the following we study whether there exists a ϵ-equilibrium . For our analysis we consider a particular metric space, and agent population model, that we describe in the next section.

4   Analysis

In this section we provide the results of the analysis of the model presented in the previous sections. Our results show that the analysis in [1,2] that was obtained under the assumption of a continuous agent population extends to the case of a discrete agent population.

For our analysis, we use the following notation. Given a real-valued function $f : \mathcal{M} \mapsto R$ on a metric space $\mathcal{M}$, we define the support $supp(f)$ by

$$supp(f) = \bar{A}$$

where $A = \{x \in \mathcal{M}|f(x) \neq 0\}$, and $\bar{A}$ is the closure of A.

Given real-valued function $f : \mathcal{M} \mapsto R$ on a metric space $\mathcal{M}$, we say that f is symmetric with respect to $y \in \mathcal{M}$ if for $x, x' \in \mathcal{M}$ such that

$$d(x, y) = d(x', y),$$

we have that

$$f(x) = f(x').$$

4.1   Content Space $\mathcal{R}$ and Discrete Agent Population Model

For our analysis we consider a metric space that has a particular structure. More precisely, we consider a one dimensional metric space with the torus metric. The reason for using this structure is that it simplifies the analysis and allows us to obtain simple expressions for our results, that can easily been interpreted.

More formally, we consider in the following one-dimensional metric space for our analysis. The metric space is given by an interval $\mathcal{R} = [-L, L) \in R$, $0 < L$, with the torus metric, i.e. the distance between two points $x, y \in \mathcal{R}$ is given by

$$d(x, y) = ||x - y|| = \min\{|x - y|, 2L - |x - y|\},$$

where $|x|$ is the absolute value of $x \in (-\infty, \infty)$. Note that we have that $||x - y|| \leq L$, $x, y \in \mathcal{R}$.

Using the torus metric for the content space $\mathcal{R}$ eliminates "border effects", in the sense are no points that have a "special" position as it would be for example the case if we would an interval $[-L, L]$ as the content space. This simplifies the analysis, and leads to simpler expressions for our results.

We assume that there exists a finite number of agents that exclusively produce content, as well as a finite number of agents that exclusively consume content. More precisely, we assume that there are $K^{(d)}$ agents that consume content which are "uniformly distributed" on $\mathcal{R}$ with distance

$$\delta_d = \frac{2L}{K^{(d)}},$$

and we have that $\mathcal{A}^{(d)} = \{y_1, ..., y_{K^{(d)}}\}$ where

$$y_{k+1} = y_k + \delta_d, \quad k = 1, ..., K^{(d)} - 1.$$

Similarly, we assume that there are $K^{(s)}$ agents that produce content which are "uniformly distributed" with distance

$$\delta_s = \frac{2L}{K^{(s)}},$$

over $\mathcal{R}$, and we have that $\mathcal{A}^{(s)} = \{y_1, ..., y_{K^{(s)}}\}$ where

$$y_{k+1} = y_k + \delta_s, \quad k = 1, ..., K^{(s)} - 1.$$

While we consider here the case where there exists a set of content producers, and a set content consumers, the results in this paper can easily be extended to the case where each agent both produces and consumes content. The results obtained in this paper also hold for this case, requiring only notational changes in the proofs.

Finally, we make the following assumption for the function f and g are used in Eq. (1) and Eq. (2) to define the agents' interest and ability to produce content.

Assumption 1. *The function $f : [0, L] \mapsto [0, 1]$ is strictly decreasing and three times continuously differentiable on $[0, L]$, the first three derivatives are bounded first derivative on $[0, L]$, and we have that $f'(0) < 0$. Furthermore, the function f is locally strictly concave, i.e. there exists a constant b, $0 < b \leq L$, such that $f''(x) < 0$, $x \in [0, b]$. The function $g : [0, L] \mapsto [0, 1]$ is non-increasing on $[0, L]$, and strictly concave and twice continuously differentiable on its support $supp(g)$ with $g(0) > 0$ and $g'(0) = 0$.*

These assumptions on the functions f and g are a technical assumptions used in the proofs of our results presented in the next section.

4.2   Community Structure $\mathcal{C}_\delta(L_C)$

In our analysis, we consider a particular class $\mathcal{C}_\delta(L_C)$ of community structures that are defined as follows.

Definition 1. *Let L_C be such that*

$$0 < 2L_C < \min\{b, L\},$$

where b is the constant of Assumption 1. The class $\mathcal{C}_\delta(L_C)$ then consists of all community structures

$$\left(\mathcal{C}, \{\alpha_C(y)\}_{y \in \mathcal{A}^{(d)}}, \{\beta_C(\cdot|y)\}_{y \in \mathcal{A}^{(s)}}\right)$$

where

$$\mathcal{C} = \{C_k\}_{k=1,\ldots,K}$$

that are defined on a set of agents $\mathcal{A}^{(d)}$ with distance δ_d such that

$$0 < \delta_d < \delta,$$

and set of agents $\mathcal{A}^{(s)}$ with distance δ_s such that

$$0 < \delta_s < \delta,$$

and have the following properties.
There exists set a $\{I_{C_k}\}_{k=1,\ldots,K}$ of mutually non-overlapping intervals in $\mathcal{R}$ of length $2L_C$, i.e we have that

1) $I_{C_k} \cap I_{C_{k'}} = \emptyset,\ k \neq k'$,
2) $\cup_{k=1,\ldots,K} I_{C_k} = \mathcal{R}$,
3) $|I_{C_k}| = 2L_C,\ k = 1,\ldots,K$,

such that the community $C_k = (C_d, C_s) \in \mathcal{C} = \{C_k\}_{k=1,\ldots,K}$, is given by

$$C_d = \mathcal{A}^{(d)} \cap I_{C_k}$$

and

$$C_s = \mathcal{A}^{(s)} \cap I_{C_k}.$$

Furthermore, for $C = (C_d, C_s) \in \mathcal{C}\{C_k\}_{k=1,\ldots,K}$ we have that

a) $\alpha_C(y) = E_p,\ y \in C_d$, *and*
b) $\beta_C(x|y) = E_q \delta(x^*(y) - x),\ y \in C_s, x \in \mathcal{R}$, *where*

$$x^*(y) = \arg\max_{x \in \mathcal{R}} [q(x|y)P_C(x)] = \arg\max_{x \in \mathcal{R}} \left[q(x|y)E_p \sum_{y \in C_d} p(x|y) \right]$$

and $\delta(\cdot)$ is the Dirac delta function.

In the next section, we show that there always exists a class $\mathcal{C}_\delta(L_C)$ of community structures such that all community structures $\mathcal{S} \in \mathcal{C}_\delta(L_C)$ are a $\epsilon-$equilibrium.

4.3 Existence of a ϵ−Equilibrium

Our first results shows that there always exists a ϵ−equilibrium.

Proposition 1. *Suppose that* $f(0)g(0) - c > 0$, *then there exists a* L_C,

$$0 < 2L_C < \min\{b, L\},$$

where b is the constant of Assumption 1, such that the following is true. For every $\epsilon > 0$ *there exists a* $\delta > 0$ *such that all discrete interval community structures* $\mathcal{S}^* \in \mathcal{C}_\delta(L_C)$ *are a* ϵ−*equilibrium.*

Proposition 1 states that there always exists a ϵ−equilibrium $\mathcal{S}^*$ given that distance δ_d and δ_s of the agents sets $\mathcal{A}^{(d)}$ and $\mathcal{A}^{(s)}$, are small enough. In addition, from the definition of the class $\mathcal{C}_\delta(L_C)$ we have that under a Nash equilibrium given by Proposition 1 a content producer $y \in C_{k,s}$ in community $C_k = (C_{k,d}, C_{k,s})$ focuses on producing a single type of content given by $x^*(y)$. This result is interesting as experimental results suggest that this property indeed holds in real-life information networks (see discussion in Sect. 2).

In the following we characterize in more details the properties of a ϵ−equilibrium as given in Proposition 1.

4.4 Optimal Content Production

Proposition 1 states that under the ϵ−equilibrium as given in Proposition 1, each agent $y \in \mathcal{A}^{(s)}$ produces a single content type $x^*(y)$. In this subsection we characterize in more details the function $x^*(y)$ for a given community $C = (C_d, C_s)$ in a ϵ−equilibrium. We have the following result.

Proposition 2. *Let* $\mathcal{C}_\delta(L_C)$ *be a class of discrete interval community structures with distance* δ *as given by Proposition 1, i.e. we have that all community structures* $\mathcal{S}^* \in \mathcal{C}_\delta(L_C)$ *are a* ϵ−*equilibrium. Then for every* Δ_{x^*}, $0 < \Delta_{x^*} < L_C$, *there exists a class* $\mathcal{C}_{\delta_0}(L_C) \subseteq \mathcal{C}_\delta(L_C)$, $0 < \delta_0 \le \delta$, *of discrete interval community structures with distance* δ_0 *such that for all community structures*

$$\left(\mathcal{C}, \{\alpha_{\mathcal{C}}^*(y)\}_{y \in \mathcal{A}^{(d)}}, \{\beta_{\mathcal{C}}^*(\cdot|y)\}_{y \in \mathcal{A}^{(s)}}\right) \in \mathcal{C}_{\delta_0}(L_C)$$

the following is true. Given a community $C = (C_d, C_s) \in \mathcal{C}$, *let the interval* $I_C = [mid(I_C) - L_C, mid(I_C) + L_C] \subset \mathcal{R}$ *as given in Proposition 1, i.e. we have that*

$$C_d = \mathcal{A}^{(d)} \cap I_C \text{ and } C_s = \mathcal{A}^{(s)} \cap I_C.$$

Then the solution $x_\delta^*(y)$ *to the optimization problem*

$$x^*(y) = \arg\max_{x \in \mathcal{R}} q(x|y) P_C(x), \qquad y \in I_C,$$

where

$$P_C(x) = E_p \sum_{y \in C_d} p(x|y),$$

has the properties that

(a) there exists a unique optimal solution $x^*(y)$, $y \in I_C$.

(b) for $y \in [mid(I_C) - L_C, mid(I_C) - \Delta_{x^*}]$, we have that

$$x^*(y) \in (y, mid(I_C)) \cap supp(q(\cdot|y)).$$

(c) for $y \in [mid(I_C) + \Delta_{x^*}, mid(I_C) + L_C]$, we have that

$$x^*(y) \in (mid(I_C), y) \cap supp(q(\cdot|y)).$$

(d) the function $x^*(y)$ is strictly increasing and differentiable on

$$\tilde{I}_C = [mid(I_C) - L_C, mid(I_C) - \Delta_{x^*}] \cup [mid(I_C) + \Delta_{x^*}, mid(I_C) + L_C].$$

Proposition 2 states that the function $x^*(y)$ is strictly increasing on $\tilde{I}_C$. This result implies that two different agents $y, y' \in \tilde{I}_C$, $y \neq y'$, produce different types of content, i.e. we have that $x^*(y) \neq x^*(y')$. This result is interesting as it states that under each agent y in $\tilde{I}_C$ produces a unique content type $x^*(y)$, i.e. we have that the content type $x^*(y)$ is not produced by any other agent in $\tilde{I}_C$.

Another interesting aspect of Proposition 2 is that Property (b) and Property (c) state that agents $y \in \tilde{I}_C$ produce content that is closer to the center of interest $mid(I_C)$ of the community C than their center of interest y. To get a more detailed understanding of how agents adapt the type of content that they produce towards the center of interest $mid(I_C)$ of the community C, we next study the function $\Delta^*(y)$ given by

$$\Delta^*(y) = ||y - x^*(y)||, \qquad y \in I_C.$$

The function $\Delta^*(y)$ characterizes the absolute value of the "displacement" of the optimal content $x^*(y)$ that agent y produces, and content y that the agent is best at producing which is equal to content type y. Or in other words, the function $\Delta^*(y)$ characterizes by how much an agent y adapts its content $x^*(y)$ towards the center of interest of the community C, i.e. by how much agent y produces content $x^*(y)$ that is closer to the center of interest $mid(I_C)$ of the community than its own center of interest.

We have the following result for the function $\Delta^*(y)$.

Proposition 3. *Let $\mathcal{C}_\delta(L_C)$ be a class of discrete interval community structures with distance δ as given by Proposition 1, i.e. we have that all community structures $\mathcal{S}^* \in \mathcal{C}_\delta(L_C)$ are a ϵ-equilibrium. Then for every Δ_{x^*}, $0 < \Delta_{x^*} < L_C$, there exists a class $\mathcal{C}_{\delta_0}(L_C) \subseteq \mathcal{C}_\delta(L_C)$, $0 < \delta_0 \leq \delta$, of discrete interval community structures with distance δ_0 such that for all community structures*

$$\left(C, \{\alpha_C^*(y)\}_{y \in \mathcal{A}^{(d)}}, \{\beta_C^*(\cdot|y)\}_{y \in \mathcal{A}^{(s)}}\right) \in \mathcal{C}_{\delta_0}(L_C)$$

the following is true. Given a community $C = (C_d, C_s) \in \mathcal{C}$ and the corresponding interval $I_C = [mid(I_C) - L_C, mid(I_C) + L_C)$, the function $\Delta^*(y)$ given by

$$\Delta^*(y) = ||y - x^*(y)||, \qquad y \in I_C,$$

is strictly decreasing and differentiable on $[mid(I_C) - L_C, mid(I_C) - \Delta_{x^*}]$, and strictly increasing and differentiable on $[mid(I_C) + \Delta_{x^*}, mid(I_C)] + L_C]$.

Proposition 3 states that the function $\Delta^*(y)$ is strictly decreasing and differentiable on $[mid(I_C) - L_C, mid(I_C) - \Delta_{x^*}]$, and strictly increasing and differentiable on $[mid(I_C) + \Delta_{x^*}, mid(I_C)] + L_C]$. This implies that the further away an agent is from the center of interest $mid(I_C)$, the more it will "adapt" the content it produces towards to the center $mid(I_C)$ of the interval I_C, i.e. the larger $\Delta^*(y)$ will be. In addition, this result implies that the further away an agent is from the center of interest $mid(I_C)$, the lower the quality is the content that the agents produces. To see this, recall that by Assumption 1, g is decreasing on $supp(g)$ and we have that the quality of the content that agent y produces is given by

$$q(x^*(y)|y) = g(||y - x^*(y)||) = g(\Delta^*(y)).$$

4.5   Properties of the Content Demand Function $P_C(x)$

We next characterize the properties of the content demand function $P_C(x)$ given by

$$P_C(x) = \sum_{y \in I_{\delta,C}} \alpha_C^*(y)p(x|y) = E_p \sum_{y \in I_{\delta,C}} p(x|y), \qquad x \in \mathcal{R}, \qquad (5)$$

of a discrete interval community $C = (C_d, C_s)$ under ϵ−equilibrium as given in Proposition 1. The demand function captures the "interest" in content on topic x within the community C. We have the following result.

Proposition 4. *Let $C_\delta(L_C)$ be a class of discrete interval community structures with distance δ as given by Proposition 1, i.e. we have that all community structures $S^* \in C_\delta(L_C)$ are a ϵ−equilibrium. Then for every Δ_P, $0 < \Delta_P < L_C$, there exists a class $C_{\delta_0}(L_C) \subseteq C_\delta(L_C)$, $0 < \delta_0 \leq \delta$, of discrete interval community structures with distance δ_0 such that for all community structures*

$$\left(C, \{\alpha_C^*(y)\}_{y \in A^{(d)}}, \{\beta_C^*(\cdot|y)\}_{y \in A^{(s)}}\right) \in C_{\delta_0}(L_C)$$

the following is true. Given a community $C = (C_d, C_s) \in C$ and the corresponding interval $I_C = [mid(I_C) - L_C, mid(I_C) + L_C)$, the demand function $P_C(x)$ given by Eq. (5) is strictly increasing on the interval $[mid(I_C) - L_C, mid(I_C) - \Delta_P]$, and strictly decreasing on the interval $[mid(I_C) + \Delta_P, mid(I_C) + L_C)$.

Note that Proposition 4 implies that

$$\arg\max_{x \in \mathcal{R}} P_C(x) \in [mid(I_C) - \Delta_P, mid(I_C) + \Delta_P],$$

i.e. the most popular content is close to the center of interest of the community C. Furthermore, we have that the further away a content type x is from the center of interest of the community, the less popular it is.

4.6   Properties of the Content Supply Function $Q_C^*(x)$

Next characterize the properties of the content demand function $Q_C^*(x)$ given by

$$Q_C^*(x) = E_q \sum_{y \in C_s} \delta\big(x - x^*(y)\big) q(x^*(y)|y). \tag{6}$$

of a discrete interval community $C = (C_d, C_s)$ under $\epsilon-$equilibrium as given in Proposition 1 where $x^*(y) = \arg\max_{x \in \mathcal{R}} q(x|y) P_C(x)$, and $\delta(\cdot)$ is the Dirac delta function. The supply function $Q_C^*(x)$ captures the overall rate with which relevant content on topic x is produced within the community C. We have the following result.

Proposition 5. *Let $\mathcal{C}_\delta(L_C)$ be a class of discrete interval community structures with distance δ as given by Proposition 1, i.e. we have that all community structures $\mathcal{S}^* \in \mathcal{C}_\delta(L_C)$ are a $\epsilon-$equilibrium. Then there exists a class $\mathcal{C}_{\delta_0}(L_C) \subseteq \mathcal{C}_\delta(L_C)$, $0 < \delta_0 \leq \delta$, of discrete interval community structures with distance δ_0 such that for all community structures*

$$\big(\mathcal{C}, \{\alpha_C^*(y)\}_{y \in \mathcal{A}^{(d)}}, \{\beta_C^*(\cdot|y)\}_{y \in \mathcal{A}^{(s)}}\big) \in \mathcal{C}_{\delta_0}(L_C)$$

the following is true. Given a community $C = (C_d, C_s) \in \mathcal{C}$ and the corresponding interval $I_C = [mid(I_C) - L_C, mid(I_C) + L_C)$, the content supply function $Q_C^*(x)$, $x \in \mathcal{R}$, as given by Eq. (6) is such that

$$supp(Q_C^*(\cdot)) \subseteq [mid(I_C) - L_C^*, mid(I_C) + L_C^*]$$

where $0 < L_C^* < L_C$.

Proposition 5 states that

$$supp(Q_C^*(\cdot)) \subseteq [mid(I_C) - L_C^*, mid(I_C) + L_C^*]$$

where $0 < L_C^* < L_C$. This result implies that the content type that is being produced by agents in the community C is a strict subset of the interval I_C. As a result, there is no overlap in the content produced in different communities under a $\epsilon-$equilibrium as given by Proposition 1.

4.7   Properties of the Utility Function $U_C^{(d)}(y)$ and $U_C^{(s)}(y)$

Finally, we study the properties of the utility rate function for content consumption $U_C^{(d)}(y)$, and the utility rate function for content production $U_C^{(s)}(y)$ for an interval community $C = (C_d, C_s)$ under a $\epsilon-$equilibrium as given by Proposition 1.

We first study the properties of the utility rates for content consumption $U_C^{(d)}(y)$ for an interval community $C = (C_d, C_s)$ under a $\epsilon-$equilibrium as given by Proposition 1.

Proposition 6.   *Let $\mathcal{C}_\delta(L_C)$ be a class of discrete interval community structures with distance δ as given by Proposition 1, i.e. we have that all community structures $\mathcal{S}^* \in \mathcal{C}_\delta(L_C)$ are a ϵ−equilibrium. Then for every Δ_U, $0 < \Delta_U < L_C$, there exists a class $\mathcal{C}_{\delta_0}(L_C) \subseteq \mathcal{C}_\delta(L_C)$, $0 < \delta_0 \le \delta$, of discrete interval community structures with distance δ_0 such that for all community structures*

$$\left(\mathcal{C}, \{\alpha^*_\mathcal{C}(y)\}_{y \in \mathcal{A}^{(d)}}, \{\beta^*_\mathcal{C}(\cdot|y)\}_{y \in \mathcal{A}^{(s)}}\right) \in \mathcal{C}_{\delta_0}(L_C)$$

the following is true. Given a community $C = (C_d, C_s) \in \mathcal{C}$ and the corresponding interval $I_C = [mid(I_C) - L_C, mid(I_C) + L_C)$, the utility rate function for content consumption $U_C^{(d)}(y)$ given by

$$U_C^{(d)}(y) = E_p E_q \sum_{z \in C_s} \left[p(x^*(z)|y) q(x^*(z)|z) - c \right], \qquad y \in C_d,$$

has the following properties.

a) *For $y, y' \in C_d \cap [mid(I_C) - L_C, mid(I_C) - \Delta_U]$ such that $y > y'$, we have that $U_C^{(d)}(y) > U_C^{(d)}(y')$.*
b) *For $y, y' \in C_d \cap [mid(I_C) + \Delta_U, mid(I_C) + L_C]$ such that $y < y'$, we have that $U_C^{(d)}(y) > U_C^{(d)}(y')$.*

Proposition 6 states that the closer an agent $y \in C_d$ is to the center of interest of the community, the higher a higher utility rate it receives. This is an interesting result as it suggest that the utility rate might can be used to rank (order) agents in an information community based on how close they are to the center $mid(I_C)$ of the community I_C.

We next study the properties of the utility rates for content production $U_C^{(s)}(y)$ for an discrete interval community $C = (C_d, C_s)$ under a ϵ−equilibrium as given by Proposition 1.

Proposition 7.   *Let $\mathcal{C}_\delta(L_C)$ be a class of discrete interval community structures with distance δ as given by Proposition 1 such that all community structures $\mathcal{S}^* \in \mathcal{C}_\delta(L_C)$ are a ϵ−equilibrium. Then for every Δ_U, $0 < \Delta_U < L_C$, there exists a class $\mathcal{C}_{\delta_0}(L_C) \subseteq \mathcal{C}_\delta(L_C)$, $0 < \delta_0 \le \delta$, of discrete interval community structures with distance δ_0 such that for all community structures*

$$\left(\mathcal{C}, \{\alpha^*_\mathcal{C}(y)\}_{y \in \mathcal{A}^{(d)}}, \{\beta^*_\mathcal{C}(\cdot|y)\}_{y \in \mathcal{A}^{(s)}}\right) \in \mathcal{C}_{\delta_0}(L_C)$$

the following is true. Given a community $C = (C_d, C_s) \in \mathcal{C}$ and the corresponding interval $I_C = [mid(I_C) - L_C, mid(I_C) + L_C)$, the utility rate function for content production $U_C^{(s)}(y)$ given by

$$U_C^{(s)}(y) = E_p E_q \sum_{z \in C_d} \left[q(x^*(y)|y) p(x^*(y)|z) - c \right], \qquad y \in C_s,$$

has the following properties.

a) *For* $y, y' \in C_s \cap [mid(I_C) - L_C, mid(I_C) - \Delta_U]$ *such that* $y > y'$, *we have that* $U_C^{(s)}(y) > U_C^{(s)}(y')$.

b) *For* $y, y' \in C_s \cap [mid(I_C) + \Delta_U, mid(I_C) + L_C]$ *such that* $y < y'$, *we have that* $U_C^{(s)}(y) > U_C^{(s)}(y')$.

Similar to Proposition 6, Proposition 7 states that the closer an agent $y \in C_d$ is to the center of interest of the community, the higher a higher utility rate it receives. Again this result suggest that the utility rate might can be used to rank (order) agents in an information community based on how close they are to the center $mid(I_C)$ of the community I_C.

5   Conclusions

In this paper we generalized the results of [1, 2] that were obtained for the case of a continuous agent model, to the case of a discrete agent population model. An interesting aspect of the obtained result is that they indeed provide insights into properties of communities in real-life information networks. Due to space constraints, we refer to [1–3] for a more detailed discussion.

In ongoing research we use the model and results presented in this paper to study the connectivity (graph structure) of how agents interact with each other, and forward content, in an information community. In addition use the model and results presented in this paper to study a novel class of local algorithms to detect (information) communities in social networks.

References

1. Marbach, P.: The structure of communities in information networks. In: Information Theory and Applications (ITA) Workshop (2016). https://doi.org/10.1109/ITA.2016.7888181
2. Marbach, P.: Modeling and analysis of information communities. http://arXiv.org/abs/1511.08904
3. Marbach, P.: Community Structures in Information Networks for a Discrete Agent Population. http://arXiv.org/abs/2004.05708
4. Bonato, A., Janssen, J., Pralat, P.: Geometric protean graphs. CoRR, abs/1111.0207 (2011)
5. Leskovec, J., Chakrabarti, D., Kleinberg, J., Faloutsos, C., Ghahramani, Z.: Kronecker graphs: an approach to modeling networks. J. Mach. Learn. Res. **11**, 985–1042 (2010)
6. Zadeh, R.B., Goel, A., Munagala, K., Sharma, A.: On the precision of social and information networks. In: Proceedings of the First ACM Conference on Online Social Networks, pp. 63–74. ACM (2013)
7. Hegde, N., Massoulié, L., Viennot, L.: Self-organizing flows in social networks. In: Moscibroda, T., Rescigno, A.A. (eds.) SIROCCO 2013. LNCS, vol. 8179, pp. 116–128. Springer, Cham (2013). https://doi.org/10.1007/978-3-319-03578-9_10

Author Index

Aksoy, Sinan 1
Antelmi, Alessia 36
Ardickas, Daumilas 82
Arendt, Dustin 1

Barla, Annalisa 145
Bloznelis, Mindaugas 68, 82, 124
Bonato, Anthony 111, 135

Cordasco, Gennaro 36
Cranston, Daniel W. 111

Firoz, Jesun 1

Garbarino, Davide 145
Gracar, Peter 96

Heydenreich, Markus 96
Huggan, Melissa A. 111

Jaworski, Jerzy 124
Jenkins, Louis 1
Joslyn, Cliff A. 1

Kamiński, Bogumił 52
Karjalainen, Joona 68

Leskelä, Lasse 68

Marbach, Peter 158
Marbach, Trent 111
Meger, Erin 135
Mönch, Christian 96
Mörters, Peter 96
Mutharasan, Raja 111

Praggastis, Brenda 1
Prałat, Paweł 52
Prokhorenkova, Liudmila 16
Purvine, Emilie 1

Rybarczyk, Katarzyna 124

Samosvat, Egor 16
Spagnuolo, Carmine 36
Szufel, Przemysław 36

Théberge, François 52
Tozzo, Veronica 145

van der Hoorn, Pim 16
Vian, Andrea 145

Zalewski, Marcin 1

Printed in the United States
By Bookmasters